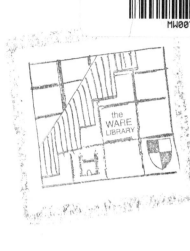

THE HARVEY LECTURES

From the engraving by Jacobus Houbraken

WILLIAM HARVEY

BORN APRIL 1, 1578–DIED JUNE 3, 1657

THE HARVEY LECTURES

DELIVERED UNDER THE AUSPICES OF

The HARVEY SOCIETY of NEW YORK

2006–2007

———

BY

KATHRYN V. ANDERSON TOM A. RAPOPORT
SUSAN K. DUTCHER GREGORY L. VERDINE
MICHAEL E. GREENBERG LEONARD I. ZON
MICHAEL KARIN

SERIES 102

2008

A JOHN WILEY & SONS, INC., PUBLICATION

For general information on our other products and services or for technical support, please contact our Customer Care Department within the United States at (800) 762-2974, outside the United States at (317) 572-3993 or fax (317) 572-4002.

Wiley also publishes its books in a variety of electronic formats. Some content that appears in print may not be available in electronic formats. For more information about Wiley products, visit our web site at www.wiley.com.

Library of Congress Cataloging-in-Publication Data

ISBN 978-0-470-59137-6

Printed in the United States of America

10   9   8   7   6   5   4   3   2   1

CONTENTS

CONSTITUTION OF THE HARVEY SOCIETY   vii

BY-LAWS OF THE HARVEY SOCIETY, INC.   ix

OFFICERS OF THE HARVEY SOCIETY   xv

CORPORATE SPONSORS   xvi

PREFACE: A BRIEF HISTORY OF THE HARVEY SOCIETY, NEW YORK   xvii

HARVEY LECTURES 2006–2007

DRUGGING THE "UNDRUGGABLE"   1
   Gregory L. Verdine, Ph.D., Erving Professor of Chemistry,
   Department of Stem Cell and Regenerative Biology,
   Chemistry and Chemical Biology, and Molecular and
   Cellular Biology, Harvard University, Cambridge,
   Massachusetts; Director, Chemical Biology Initiative and
   Program in Cancer Chemical Biology, Dana-Farber Cancer
   Institute, Boston, Massachusetts

BASAL BODIES: THEIR ROLES IN GENERATING ASYMMETRY   17
   Susan K. Dutcher, Ph.D., Professor of Genetics, Department
   of Genetics, Washington University School of Medicine,
   St. Louis, Missouri

PROTEIN TRANSPORT IN AND OUT OF THE ENDOPLASMIC RETICULUM . . .   51
   Tom A. Rapoport, Ph.D., Professor, Howard Hughes Medical
   Institute and Department of Cell Biology, Harvard Medical
   School, Boston, Massachusetts

SIGNALING NETWORKS THAT CONTROL SYNAPSE DEVELOPMENT
AND COGNITIVE FUNCTION 73
    Michael E. Greenberg, Ph.D., Professor of Neurology,
    Professor of Neurobiology, Children's Hospital Boston,
    Program in Neurobiology, Harvard Medical School,
    Department of Neurobiology, Boston, Massachusetts

CILIA AND HEDGEHOG SIGNALING IN THE MOUSE EMBRYO 103
    Kathryn V. Anderson, Ph.D., Program Chair, Developmental
    Biology Program, Sloan-Kettering Institute, New York,
    New York

DERIVATION OF ADULT STEM CELLS DURING EMBRYOGENESIS 117
    Leonard I. Zon, M.D., Grousbeck Professor of Pediatrics,
    Children's Hospital Boston, Howard Hughes Medical
    Institute, Boston, Massachusetts

TRACKING THE ROAD FROM INFLAMMATION TO CANCER:
THE CRITICAL ROLE OF IκB KINASE (IKK) 133
    Michael Karin, Ph.D., Distinguished Professor of
    Pharmacology, Laboratory of gene Regulation and Signal
    Transduction, Departments of Pharmacology and Pathology
    Cancer Center, School of Medicine, University of California,
    San Diego, La Jolla, California

FORMER OFFICERS OF THE HARVEY SOCIETY.................... 153

CUMULATIVE AUTHOR INDEX 169

ACTIVE MEMBERS 179

CONSTITUTION OF THE HARVEY SOCIETY*

I

This Society shall be named the Harvey Society.

II

The object of this Society shall be the diffusion of scientific knowledge in selected chapters in anatomy, physiology, pathology, bacteriology, pharmacology, and physiological and pathological chemistry, through the medium of public lectures by men and women who are workers in the subjects presented.

III

The members of the Society shall constitute two classes: Active and Honorary members. Active members shall be workers in the medical or biological sciences, residing in the metropolitan New York area, who have personally contributed to the advancement of these sciences. Active members who leave New York to reside elsewhere may retain their membership. Honorary members shall be those who have delivered lectures before the Society and who are not Active members. Honorary members shall not be eligible to office, nor shall they be entitled to a vote.

Active member shall be elected by ballot. They shall be nominated to the Executive Committee and the names of the nominees shall accompany the notice of the meeting at which the vote for their election will be taken.

---

*The Constitution is reprinted here for historical interest only; its essential features have been included in the Articles of Incorporation and By-Laws.

IV

The management of the Society shall be vested in an Executive Committee to consist of a President, a Vice-President, a Secretary, a Treasurer, and three other members, these officers to be elected by ballot at each annual meeting of the Society to serve one year.

V

The Annual Meeting of the Society shall be held at a stated date in January of each year at a time and place to be determined by the Executive Committee. Special meetings may be held at such times and places as the Executive Committee may determine. At all meetings ten members shall constitute a quorum.

VI

Changes in the Constitution may be made at any meeting of the Society by a majority vote of those present after previous notification to the members in writing.

BY-LAWS OF THE HARVEY SOCIETY, INC.

ARTICLE I

Name and Purposes of the Society

SECTION 1. The name of the Society as recorded in the Constitution at the time of its founding in 1905 was The Harvey Society. In 1955, it was incorporated into the State of New York as The Harvey Society, Inc.

SECTION 2. The purposes for which this Society is formed are those set forth in its original Constitution and modified in its Certificate of Incorporation as from time to time amended. The purposes of the Society shall be to foster the diffusion of scientific knowledge in selected chapters of the biological sciences and related areas of knowledge through the medium of public delivery and printed publication of lectures by men and women who are workers in the subjects presented, and to promote the development of these sciences.

It is not organized for pecuniary profit, and no part of the net earnings, contributions, or other corporate funds of the Society shall inure to the benefit of any private member or individual, and no substantial part of its activities shall be carrying on propaganda, or otherwise attempting, to influence legislation.

ARTICLE II

Offices of the Society

SECTION 1. The main office and place of business of the Society shall be in the City and County of New York. The Board of Directors may designate additional offices.

ARTICLE III

Members

SECTION 1. The members of the Society shall consist of the incorporators, members of the hitherto unincorporated Harvey Society, and persons elected from time to time. The members of the Society shall constitute two classes: Active and Honorary Members. Active members shall be individuals with either the Ph.D. or the M.D. degree or its equivalent, residing or carrying on a major part of their work in the New York metropolitan area at the time of their election, who are personally making original contributions to the literature of the medical or biological sciences. Honorary members shall be those who have delivered a lecture before the Society and who are not Active members. Honorary members shall be exempted from the payment of dues. Active members who have remained in good standing for 35 years or who have reached the age of 65 and have remained in good standing for 25 years shall be designated Life members. They shall retain all the privileges of their class of membership without further payment of dues. Honorary members shall not be eligible to office, nor shall they be entitled to participate by voting in the affairs of the society. Volumes of The Harvey Lectures will be circulated only to Active members. Life members will be offered the opportunity to purchase The Harvey Lectures at the cost of the volume. Honorary members will receive only the volume containing their lecture. New Active members shall be nominated in writing to the Board of Directors by an Active member and seconded by another Active member. They shall be elected at the Annual Meeting of the Society by a vote of the majority of the Active members present at the meeting. Members who leave New York to reside elsewhere may retain their membership. Active members who have given a Harvey Lecture and who have moved out of the New York metropolitan area may, if they wish, become Honorary members. Membership in the Society shall terminate on the death, resignation, or removal of the member.

SECTION 2. Members may be suspended or expelled from the Society by he vote of a majority of the members present at any meeting of members at which a quorum is present, for refusing or failing to comply with the By-Laws, or for other good and sufficient cause.

SECTION 3. Members may resign from the Society by written declaration, which shall take effect upon the filling thereof with the Secretary.

ARTICLE IV

Meetings of the Members of the Society

SECTION 1. The Society shall hold its annual meeting to Active members for the election of officers and directors, and for the transaction of such other business as may come before the meeting in the month of January or February in each year, at a place within the City of New York, and on a date and at an hour to be specified in the notice of such meeting.

SECTION 2. Special meetings of members shall be called by the Secretary upon the request of the President or Vice-President or of the Board of Directors, or on written request of twenty-five of the Active members.

SECTION 3. Notice of all meeting of Active members shall be mailed or delivered personally to each member not less than ten nor more than sixty days before the meeting. Like notice shall be given with respect to lectures.

SECTION 4. At all meetings of Active members of the Society ten Active members, present in person, shall constitute a quorum, but less than a quorum shall have power to adjourn from time to time until a quorum be present.

ARTICLE V

Board of Directors .

SECTION 1. The number of direction constituting The Board of Directors shall be seven: the President, the Vice-President, the Secretary, and the Treasurer of the Society, and the four members of the Council. The number of directors may be increased or reduced by amendments of the By-Laws as hereinafter provided, within the maximum and minimum numbers fixed in the Certificate of Incorporation or any amendment thereto.

SECTION 2. The Board of Directors shall hold an annual meeting shortly before the annual meeting of the Society.

Special meetings of the Board of Directors shall be called at any time by the Secretary upon the request of the President or Vice-President or of one-fourth of the directors then in office.

SECTION 3. Notice of all regular annual meetings of the Board shall be given to each director at least seven days before the meeting and notice of special meetings, at least one day before. Meetings may be held at any place within the City of New York designated in the notice of the meeting.

SECTION 4. The Board of Directors shall have the immediate charge, management, and control of the activities and affairs of the Society, and it shall have full power, in the intervals between the annual meetings of the Active members, to do any and all things in relation to the affairs of the Society.

SECTION 5. Council members shall be elected by the members of the Society at the Annual Meeting. One Council member is elected each year to serve for three years, there being three Council members at all times. Vacancies occurring on the Council for any cause may be filled for the unexpired term by the majority vote of the directors present at any meeting at which a quorum is present. Only Active members of the Society shall be eligible for membership on the Council.

SECTION 6. A majority of the Board as from time to time constituted shall be necessary to constitute a quorum, but less than a quorum shall have power to adjourn from time to time until a quorum be present.

SECTION 7. The Board shall have power to appoint individual or corporate trustees and their successors of any or all of the property of the Society, and to confer upon them such of the powers, duties, or obligations of the directors in relation to the care, custody, or management of such property as may be deemed advisable.

SECTION 8. The directors shall present at the Annual Meeting a report, verified by the President and Treasurer, or by a majority of the directors, showing the whole amount of real and personal property owned by the Society, where located, and where and how invested, the amount and nature of the property acquired during the year immediately preceding the date of the report and the manner of the acquisition; the amount applied, appropriated, or expended during the year immediately preceding such date, and the purposes, objects, or persons to or for which such applications, appropriations, or expenditures have been made; and the names of the persons who have been admitted to membership in the Society during such year, which report shall be filed with the records of the Society and an abstract thereof entered in the minutes of the proceedings of the Annual Meeting.

ARTICLE VI

Committees

SECTION 1. The Board of Directors may appoint from time to time such committees as it deems advisable, and each such committee shall exercise such powers and perform such duties as may be conferred upon it by the Board of Directors subject to its continuing direction and control.

ARTICLE VII

Officers

SECTION 1. The officers of the Society shall consist of a President, a Vice-President, a Secretary, and a Treasurer, and such other officers as the Board of Directors may from time to time determine. All of the officers of the Society shall be members of the Board of Directors.

SECTION 2. The President shall be the chief executive officer of the Society and shall be in charge of the direction of its affairs, acting with the advice of the Board of Directors. The other offices of the Society shall have the powers and perform the duties that usually pertain to their respective offices, or as may from time to time be prescribed by the Board of Directors.

SECTION 3. The officers and the directors shall not receive, directly or indirectly, any salary or other compensation from the Society, unless authorized by the concurring vote of two-thirds of all the directors.

SECTION 4. The officers shall be elected at the Annual Meeting of the Active members. All officers shall hold office until the next Annual Meeting and until their successors are elected or until removed by vote of a majority of the directors. Vacancies occurring among the officers for any cause may be filled for the unexpired term by the majority vote of the directors present at any meeting at which a quorum is present. Officers must be Active members of the Society.

ARTICLE VIII

Fiscal Year – Seal

SECTION 1. The fiscal year of the Society shall be the calendar year.

SECTION 2. The seal of the Society shall be circular in form and shall bear the words "The Harvey Society, Inc., New York, New York, Corporate Seal."

ARTICLE IX

Amendments

SECTION 1. These By-Laws may be added to, amended, or repealed, in whole or in part, by the Active members or by the Board of Directors, in each case by a majority vote at any meeting at which a quorum is present, provided that notice of the proposed addition, amendment, or repeal has been given to each member or director, as the case may be, in the notice of such meeting.

CORPORATE SPONSORS

BD Biosciences
GE Global Research and Development
Merck Research Laboratories
Olympus America, Inc.
Pfizer Global Research and Development

PREFACE:
A BRIEF HISTORY OF THE HARVEY SOCIETY, NEW YORK

The Harvey Society was founded April 1, 1905 by 13 New York scientists and physicians, who met at 9 East 74th Street, the home of physiologist Graham Lusk. The stated purpose of the Society was to forge a "closer relationship between the purely practical side of medicine and the results of laboratory investigation." The group included Samuel J. Meltzer, William H. Park, Edward K. Dunham, James Ewing, Frederick S. Lee, Christian Herter, Simon Flexner, George B. Wallace, Theodore C. Janeway, Phoebus A. Levene, and Eugene L. Opie. The meeting was also attended by John J. Abel, a noted pharmacologist from The Johns Hopkins University School of Medicine.

Founding officers:

President	Graham Lusk
Vice-President	Simon Flexner
Secretary	George B. Wallace
Treasurer	Frederick S. Lee

The Harvey Society lectures reflect "the evolution of physiology and physiological chemistry into biochemistry and the development of molecular biology from the roots of bacteriology and biochemistry" in this century. One of the important facets of this lecture series is that they are published annually. The Harvey Society is continuing to achieve its primary goal of providing "a series of distinguished lectures in the life sciences" and, in fact, "there are a few lectures of such distinction delivered uninterruptedly over a period of more than 70 years in any other city in the world."[1]

Harvey once wrote, "My trust is in my love of truth and the candour of cultivated minds." For as long as those who are invited to deliver a Harvey Lecture heed these words, they honour the name of William Harvey, and the Harvey Society of New York has a high and noble future.[1]

---

[1] Exerpted from an article in *Perspectives in Biology and Medicine*, Summer 1978, pages 524–535, by A.G. Bearn and D.G. James.

DRUGGING THE "UNDRUGGABLE"

GREGORY L. VERDINE

Department of Stem Cell and Regenerative Biology, Chemistry and Chemical Biology, and Molecular and Cellular Biology, Harvard University, Cambridge, Massachusetts; Chemical Biology Initiative and Program in Cancer Chemical Biology, Dana-Farber Cancer Institute, Boston, Massachusetts

I. INTRODUCTION

The stunning pace of advancement in uncovering the molecular basis for human disease has not been matched by a corresponding profusion of drugs that exploit these newly gained mechanistic insights. Though not one single factor is responsible for this lag in pharmaceutical innovation, there is good reason to believe that a major factor lies in the limited toolkit of molecules available to drug discoverers today. The great majority of approved drugs and investigational agents currently undergoing human clinical testing belong to two broad molecular classes, namely small molecules and biologicals. Small molecules typically possess fewer than a hundred atoms, with a molecular mass under 1000 daltons, whereas biologicals may possess thousands of atoms and may reach masses above 100,000 daltons. The limited size of small molecules can endow them with highly beneficial pharmaceutical properties, such as the ability to penetrate broadly and deeply into the tissue of living animals, and to diffuse passively across biological hindrances such as the outer cell membrane and the blood–brain barrier. But this economy of atoms also places significant limitations on the targeting ability of small molecules. With only a limited surface area available for engagement in energetically favorable contacts with a molecular target, small molecules require engulfment by their targets in order to maximize the total contact surface area. This simple biophysical imperative underlies the well-known principle that small molecules are, for the most part, capable of targeting only those proteins that possess a deep surface involution lined with hydrophobic

1

The Harvey Lectures, Series 102, pages 1–16

amino acid side chains, a feature commonly referred to as a "hydrophobic pocket." How seriously does this restriction limit the utility of small molecules in drug discovery? While the answer to this question may never be known precisely, current estimates based on surface character analysis of proteins in the Protein Data Bank, and on empirical observations gained through high-throughput library screening, place the number of human proteins amenable to small molecule targeting at approximately 2000, a mere ~10% of the proteins encoded in the human genome (Hopkins and Groom, 2002; Russ and Lampel, 2005).

Unlike small molecules, biologicals possess large contact surface areas, and this largely frees them from restrictions on the types of target surfaces they can bind; protein : protein interaction surfaces vary widely in shape and chemical composition. The ability to select and evolve diversifiable protein scaffolds, initially monoclonal antibodies but now encompassing many different types of protein scaffolds (Binz et al., 2005; Skerra, 2007; Nuttall and Walsh, 2008; Stumpp et al., 2008), offers a virtually limitless palette of protein binders with which to target disease-relevant proteins. That said, protein therapeutics suffer from a targeting problem of their own, which imposes severe limitations on the scope of applicability: all but a few proteins lack the ability to traverse cell membranes, and there-fore, the targeting range of biologics is essentially restricted to the subset of human proteins that are expressed on the cell surface or are secreted by cells. Current estimates, again imperfect but in the right ballpark, indicate that extracellular-accessible proteins comprise another ~10% of all human proteins (Verdine and Walensky, 2007).

Assuming generously that the ~10% of proteins with hydrophobic pockets is completely distinct from the ~10% that is accessible from outside the cell, one arrives at the shocking conclusion that only a meager ~20% of all human proteins are targetable by the two well-established targeting classes of molecules, and that an overwhelming majority of all prospective targets, greater than 80%, fall into the molecular limbo of targets currently considered "undruggable." No area of therapy is left unscathed by the problem of undruggability. Cancer presents an especially compelling example, in part not only because of the urgency of the need for new treatments but also because of the vast and ever-accumulating body of human genetic and animal model data that together represent a smoking gun for the involvement of certain oncoproteins as indispensable drives of tumor cell proliferation. The majority of these are proteins that regulate transcription initiation and that participate in cell signaling via engagement of intracellular protein:protein interactions, two classes of proteins that are widely considered among the most intractable of undruggable targets. This intractability explains why K-Ras and cMyc, the two most widely activated oncoproteins in all of human cancer, today continue their lethal rampage with no impediment even remotely within sight. The enormous and ever-accumulating opportunity costs associated with the inability to target the 80% of intractable proteins continues to provide a powerful incentive for molecule discoverers to develop new ways to "drug the undruggable."

The targeting deficiencies of small molecules and biologics result from factors that are fundamental to the chemical structures of these molecules. It thus stands to reason that new classes of molecules having fundamentally new targeting properties will be required to meet the challenge presented by intractable targets. It is worth noting that nucleic acid (NA) therapeutics such as antisense oligonucleotides and siRNAs offer considerable promise to access intractable targets by interfering with the activity not of the proteins themselves, but of the RNA encoding them. NA therapeutics have proven quite challenging to develop, primarily because they rarely show adequate levels of systemic bioavailability, but efforts to overcome this hurdle are both intensive and wide ranging.

Our own efforts toward drugging undruggable targets have focused on discovering new classes of molecules that we call "synthetic biologics" because they combine the versatile surface-recognition properties of biologics with the cell permeability and synthetic manipulability of small molecules. We further decided to focus our attention on the broad goal of creating molecules that could exert their actions directly at the level of intracellular protein:protein interactions. Numerous previous attempts have been made to target protein:protein interactions through structural mimicry (Hershberger et al., 2007) or fragment-based approaches (Bartoli et al., 2007; Hajduk and Greer, 2007); thus far, these efforts have met with limited success. We reasoned that nature had already spent billions of years to arrive at certain winning solutions to the problem of targeting proteins, and therefore, instead of attempting to mimic nature, we elected to borrow directly from it. Our synthetic biologics therefore represent stripped-down versions of proteins retaining primarily the structural elements that come into direct contact with a target. This exercise in protein deconstruction is most straightforward to practice when most or all of the target contact region of a protein reside on one contiguous polypeptide chain because in such cases, the chain connectivity in the protein provides the connectivity in the peptide. Noncontiguous contact regions, on the other hand, require one to confront the difficult and largely unsolved problem of connecting the various fragmentary contact elements correctly and efficiently. The practical advantages of contiguous contact elements led us to focus on one particular element of protein secondary structure, the left-handed α-helix, a self-contained folding unit that is employed more widely than any other unit in intracellular protein:protein interactions (Henchey et al., 2008).

Past efforts at protein deconstruction have occasionally afforded the so-called "dominant negative" peptides useful for *in vitro* biochemical studies, but have rarely provided agents with robust activity in cellular assays or in animals. Why should the very same peptide sequence be potently active when incorporated into a protein but devoid of activity when disincorporated from the protein? The answer to this is straightforward: the remainder of the protein provides a scaffold that stabilizes the bioactive conformation of the incorporated peptide, and consequently, disincorporation leads to unfolding of the peptide. Unfolding of the peptide typically renders it essentially devoid of biological activity, for several well-understood reasons. Firstly, whereas the folded contact region is typically pre-organized in its bioactive conformation, the unfolded peptide exists as an enormous mixture of rapidly interconverting species. Binding of the peptide comes at the significant entropic cost of ordering an inherently disordered molecule; hence, the disincorporated peptide binds the target much more weakly than the parent protein from which it was derived. Secondly, unfolding exposes the peptide to proteolytic inactivation. The large body of structural information available on pro-teases bound to substrates and inhibitors has revealed that, regardless of mechanistic class, all of these enzymes bind their substrates in an extended conformation, with the scissile amide bond fully engaged in the active site and disengaged from folding interactions. Folding thus protects poly-peptides from proteolysis and unfolding exposes them to it. Thirdly, the amide bonds in a peptide each carry substantial dispersed charge, to which solvent water molecules are attracted and which pose a substantial impedi-ment toward passive diffusion through the cell membrane. Folding of a peptide results in the shedding of amide-bound water molecules and reduction of exposed charge in the backbone by internal charge pairing, both of which are expected to increase the probability of passive membrane diffusion. Any one of the problems caused by unfolding would be sufficient to render peptides unsuitable for most studies whose success hinges on activity in a biological setting. The combination of the three problems has had a devastating effect on the development of peptides as chemical genetic ligands and as drugs.

In considering potential solutions to the problems mentioned above, we were struck by the realization that all three – weak binding, high proteolytic susceptibility and poor cell penetration – were caused or exacerbated by unfolding, and thus, all might be addressed in one fell swoop by introducing a structural modification that restores the original bioactive conformation. We were not the first to make this realization; prior to our entry into the field, synthetic enforcement of α-helical structure had been demonstrated in several systems (Henchey et al., 2008), but to our knowledge, none of these had been demonstrated to provide the combination of high levels of helix induction, proteolytic stability and activity in cells. Some of these systems introduced functionality that seems to be incompatible, by design, with our ultimate objective of modulating intracellular protein:protein interactions; lactam bridges, for example, which would be expected to impede cell penetration, or disulfide linkages, which would be cleaved upon exposure to the reductive environment present inside cells.

Our design for an α-helix enforcement system (Schafmeister et al., 2000), now known as a helix "staple," employs only membrane-compatible hydrocarbon atoms and combines two distinct and individually powerful modes of conformational bias toward a right-handed α-helical structure. One biasing element of the staple consists of an all-hydrocarbon cross-link attached to the peptide backbone so as to straddle either one or two successive turns of the helix (i and $i+4$ or $i+7$ positions, respectively); this provides a global macrocyclic constraint that extends over the entire cross-linked segment. The second biasing element of the staple consists of α-methylation at the same α-carbon atoms that serve as attachment points for the all-hydrocarbon cross-link; this introduces a local conformational bias on the peptide chain at the site of the α-methyl, α-cross-linked amino acid against non-α-helical torsion angles (Karle and Balaram, 1990; Marshall et al., 1990). The chemical composition of the stapling system was conceived and optimized in our lab by Chris

Schafmeister, who demonstrated that the staple could be introduced into peptides in a synthetically efficient and operationally simple manner by incorporating two α-methyl,α-n-alkenyl amino acids into the peptide at the *i* and either *i*+4 or *i*+7 positions, and then performing ruthenium-mediated olefin metathesis (Grubbs, 2004) on the resin-bound peptide to close the macrocyclic hydrocarbon cross-link (Schafmeister et al., 2000). We chose olefin metathesis for the ring-closing reaction because it is one of the very few C-C bond-forming reactions that can be performed efficiently and under mild conditions on highly functionalized molecules such as peptides (Miller et al., 1996), and it is the powerful combination of these attributes that earned the developers of the reaction, Grubbs, Schrock and Chauvin, the Nobel Prize in Chemistry in 2005. In studies on model peptides derived from RNase A, Schafmeister, and later Young Woo Kim, in our lab have found that the *i,i*+4 staple is formed most effectively and gives the greatest extent of helix stabilization when both amino acids bear the *S*-stereochemical configuration and the product contains an 8-carbon cross-link (formed using two units of *S*-α-methyl,α-n-pentenylglycine, designated **S5**); the *i,i*+7 staple is optimal when the N-terminal amino acid is *R*-configurated and the C-terminal amino acid is S-configurated, with an 11-carbon cross-link [usually formed using either one unit of *R*-α-methyl,α-n-octenylglycine (**R8**) and one of **S5**; or one each of **R5** and one of **S8**)] (Schafmeister et al., 2000; Kim and Verdine, 2009). It should be noted that that prior to our studies, Grubbs, O'Leary and coworkers had reported the use of olefin metathesis on peptides containing two O-allylated *l*-serine residues to introduce *i,i*+4 cross-links into peptides, but this particular system was found to enforce the $_{3,10}$-helical conformation in solution, rather than the α-helical conformation (Blackwell et al., 2001; Boal et al., 2007).

The first application of our peptide stapling system to a biological problem was undertaken by Loren Walensky, an Oncology Fellow at the Dana-Farber Cancer Institute (DFCI) who created a vital bridge between our lab with that of Stan Korsmeyer at the Howard Hughes Medical Institute, DFCI and Harvard Medical School. The Korsmeyer lab had done seminal work on the mitochondrial events leading to the initiation of programmed cell death in human cells, work that had highlighted the importance of intracellular protein:protein interactions as mediators of apoptotic inhibition and activation (Korsmeyer, 1992; Chao and Korsmeyer, 1998; Gross et al., 1999; Danial and Korsmeyer, 2004). It occurred to us that stapled peptides might provide a unique opportunity to study and modulate apoptotic signaling in human cells. Our focus was on proteins of the Bcl-2 family, members of which reside on the cytoplasmic face of the outer mitochondrial membrane. These structurally related proteins come in three distinct functional guises: anti-apoptotic members such as Bcl-2, Bcl-$_{XL}$ and Mcl-1, which are required to activate apoptosis via permeabilization of the mitochondrial membrane; pro-apoptotic members such as Bax, Bak and Mcl-1 that serve to inhibit programmed cell death; and mediators such as Bid and Bim that act as receivers for apoptotic stimuli in the cell, and that transduce those signals through direct interactions with pro- and anti-apoptotic Bcl-2 family members (Walensky, 2006; Danial, 2007). These three families of protein constitute, in effect, a cell fate homeostat, the position of which determines whether a cell survives or undergoes programmed death. In the absence of apoptotic signals, the mediator proteins are primarily engaged with anti-apoptotic Bcl-2 family members, hence the cell survives; upon activation of apoptotic signals, the mediators then engage with pro-apoptotic Bcl-2 family members, and cell death ensues. The default position of the homeostat in normal cells is usually tipped toward survival, but a wide variety of developmental and environmental cues can reset the cell fate switch to favor death. For example, cells constantly monitor the state of their genomes, and under conditions in which genomic aberrations accumulate beyond a certain threshold level, apoptotic mediators become activated and cell death ensues. One of the most profound discoveries made by Stan Korsmeyer and his coworkers is that cancer cells can perturb the position of the cell death homeostat so as to acquire survivability despite having accumulated otherwise lethal levels of genomic aberrations. Specifically, it was found that follicular lymphoma cells contain a gene translocation results in constitutive overexpression of the *bcl-2* gene from the IgH promoter. These cells produce an abnormally large reservoir of the pro-survival Bcl-2 protein, which sequesters apoptotic mediators, thereby preventing them through mass action from productive engagement with pro-apoptic family members (Korsmeyer, 1992). Thus, Bcl-2 overexpression causes derangement of cell fate homeostat so as to favoring survival in cells that should be undergoing apoptotic death. A large body of subsequent work has extended the generality of these initial observations, leading to the current view that most, if not all, transformed cells harbor derangements in apoptotic pathways that support tumor survival and growth by subverting programmed cell death (Hanahan and Weinberg, 2000).

We reasoned that it might be possible to reset the position of the apoptotic homeostat in cancer cells that overexpress Bcl-2 by interfering with the ability of Bcl-2 to bind apoptotic mediators such as Bid and Bim. A substantial body of structural work was available to guide this effort. All pro- and anti-apoptotic Bcl-2 family members contain three signature motifs, designated BH1, BH2 and BH3 (for Bcl-2 homology 1, 2 and 3). Apoptotic mediators, though structurally related to BH1,2,3 proteins, exhibit sequence similarity only in their BH3 motif (Petros et al., 2004; Walensky, 2006; Danial, 2007). The BH3 domain of these BH3-only apoptotic mediators present the majority of the contact surface responsible for binding to BH1,2,3 proteins, and peptides comprising just the BH3 domain of mediators bind weakly but specifically to BH1,2,3 proteins. Fesik and coworkers were the first to demonstrate that a BH3 peptide from BAK binds in a shallow cleft on Bcl-$_{XL}$, and that this binding event induces the folding of the BH3 peptide from a random coil into an extended α-helical structure (Sattler et al., 1997). Subsequent structural studies have shown that the BH3 domain indeed adopts an α-helical structure when incorporated into a BH3-only protein (Petros et al., 2004), which makes it virtually certain that the α-helix is the bioactive secondary structure for all BH3 domains in proteins and in peptides derived therefrom. Additional structures of BH3 domain peptides bound to anti-apoptotic BH1,2,3 proteins have revealed that these peptides binds in a structurally related cleft on the proteins (Petros et al., 2004). However, very recent studies have introduced a new wrinkle into this structural story by showing that the binding site employed by the Bid BH3 domain to interact with and thus activate Bax, a pro-apoptotic Bcl-2 family member, is completely distinct from the conserved cleft used by BH3 domains to contact anti-apoptotic Bcl-2 family members (Gavathiotis et al., 2008).

Taking all of the aforementioned factors into account, we set out to test the notion that stapling of a BH3 domain peptide might yield a stable, cell-permeable ligand capable of displacing BH3-only proteins from pro-survival Bcl-2 family members, thereby liberating the BH3-only protein to engage pro-death members of the family and consequently activate apoptosis. We initially selected the BH3 domain from Bid for stapling because Bid appeared to be able to interact with both pro- and anti-apoptotic receptors, and this increased the likelihood that notwithstanding the unknown complexities of apoptosis biology we would be able to observe some sort of biological effect with the stapled peptide. To identify the optimal position in the 22-amino-acid Bid BH3 domain for incorporation of the staple, and the optimal type of staple, a small library of stapled peptides was constructed and individual members were assayed for various physical and functional properties such as percent helicity, solubility, Bcl-2 binding affinity, cell-permeability and so on. Based on the results from this screen, we selected the stapled peptide designated SAHB$_A$ (stapled alpha helix of BH3, isomer A) for further investigation (Walensky et al., 2004).

Circular dichroism (CD) spectra of all α-helical proteins and peptides exhibit characteristic minima at wavelengths of 208 and 222 nm, and the intensity of these as a function of concentration can be used to determine the percent helicity at a given temperature. CD analysis of the unmodified Bid BH3 domain showed that it contained less than 20% helical character at room temperature. Introduction of a single $i,i+4$ staple into the center of this peptide, thus giving SAHB$_A$, increased the helicity to a value approaching 90% (Walensky et al., 2004). Though only a five-amino-acid stretch in SAHB$_A$ comprises the stapled portion, the gain in helicity extends over nearly the entire 22 amino acids of the peptide, which provides strong evidence that the staple serves to nucleate the α-helical structure on both its N- and C-terminal flanking regions. Subsequently, studies on a diverse array of stapled peptides having different sequences and staple types have shown that the helix-nucleating influence of the staple can extend over stretches as long as ~20 amino acids, providing that they contain no helix-disrupting sequences. Resistance to thermal melting or guanidine denaturation provides another means to assess conformational stability in peptides and proteins. We found that whereas the unmodified Bid BH3 22-mer has a transition temperature for half melting (T_m) below 15°C, SAHB$_A$ exhibited a T_m of 64°C. Introduction of the staple thus had the remarkable effect of conferring on a small BH3 peptide the extreme thermal stability of a thermophilic protein.

We next assayed the effect of stapling on the binding of the Bid peptides to Bcl-2 using fluorescence polarization spectroscopy. The unmodified Bid BH3 peptide bound Bcl-2 with an equilibrium dissociation constant (K_d) of ~270 nM, while SAHB$_A$ bound with a K_d of ~40 nM, a ~7-fold improvement (Walensky et al., 2004). In the design of SAHB$_A$, we had intentionally modified only the face of the BH3 helix that in the Bid protein were directed inward toward the packed protein core, and we avoided modifying residues that projected outward from the protein surface. This being the case, the most likely source of the gain in binding affinity is from the entropic benefit of pre-organizing the peptide in its bioactive conformation. Nearly all of the stapled peptides studied in our lab to date show some gain in binding affinity, with the 7-fold effect seen with Bid on the lower end of the spectrum and the ~5000-fold gain seen with an hDM2-binding peptide being at the high end (Bernal et al., 2006). It could well be in the latter case that the staple not only pre-organizes the peptide in a helical conformation but also gains some affinity from direct contacts with hDM2.

If stapled peptides embody the target-binding domain of BH3-only proteins, then the peptides should recapitulate the binding specificity of the proteins from which they were derived. This was assessed by creating a panel of stapled BH3 peptides and measuring their binding affinities for a series of BH1,2,3 targets (Walensky et al., 2006). The stapled peptides indeed were found to exhibit specificities that closely mirrored those of their parent BH3-only proteins. For example, the stapled Bad BH3 peptide showed narrow specificity for only anti-apoptotic targets, much like the Bad protein, whereas the Bid and Bim stapled peptides exhibited the ability to target both pro- and anti-apoptotic Bcl-2 family members. Like the Bid and Bim proteins, the Bid stapled peptide did not target Mcl-1 effectively, whereas the Bim stapled peptide did. The observation that stapled Bid and Bim peptides could bind the pro-apoptotic Bax and Bax directly and specifically to activate apoptosis has helped to resolve a long-standing debate in the apoptosis community as to whether the corresponding BH3-only proteins act in a similar manner (Walensky et al., 2006). Recent structural studies by Walensky and coworkers have led to the remarkable observation that the binding site for the stapled Bid peptide – and by extension, the Bid protein – is on the opposite side of the Bad protein from the canonical BH3-binding pocket of anti-apoptotics, and this unexpected finding helps to resolve some of the confusion in the field concerning the mode of apoptosis induction by certain BH3-only proteins such as Bid and Bim (Gavathiotis et al., 2008).

Extensive work done by the Walensky lab has generated an invaluable panel of cell-permeable, bioactive stapled peptides representing virtually every known BH3 domain from a BH3-only protein, and these invaluable reagents promise to yield a treasure trove of information to elucidate the complexities of apoptosis biology (Bird et al., 2008; Pitter et al., 2008).

Stapled BH3 domain peptides may also hold promise as biological tools and human therapeutics outside of the apoptosis area. Nika Danial, in work begun while a Postdoctoral Fellow with Stan Korsmeyer and continued independently, has recently shown that the BH3-only protein Bad is a component of the mitochondrial glucokinase complex and is required to promote insulin secretion by pancreatic islet beta cells in response to exposure to elevated levels of glucose. This effect is stimulated by phosphorylation on the BH3 domain of Bad. Working in collaboration, the Danial and Walensky labs demonstrated that a phosphorylated, stapled version of the Bad BH3 domain exhibits the ability to recouple insulin secretion in response to glucose challenge in mouse Bad$^{-/-}$ islets (Danial et al., 2008). This work points to a potential therapeutic application of stapled Bad peptides in the treatment of type II diabetes.

The original Schafmeister design of stapled peptides aimed to increase the likelihood that these molecules would passively diffuse across the outer cell membrane by (1) utilizing only hydrophobic atoms in the staple, and (2) maximizing the internal hydrogen bonding of backbone amides (a consequence of maximizing α-helicity). When Loren Walensky performed our first analysis of cell permeability by a stapled peptide, namely SAHB$_A$, he found indeed that indeed the stapled peptide was cell permeable, whereas the corresponding unmodified Bid BH3 peptide was not (Walensky et al., 2004). He furthermore found that SAHB$_A$ was localized to the mitochondrial membrane, in which the molecular targets of the stapled peptide reside; on the other hand, a control peptide bearing a point mutation that significantly weakened Bcl-2 binding also permeated cells but did not show mitochondrial localization. The fortunate surprise came when Walensky observed that the stapled peptide did not diffuse passively across the cell membrane but instead was actively imported into cells through an energy-dependent manner mediated by endocytic vesicles (Walensky et al., 2004). Also extremely important was the observation that the import vesicles do not seem to fuse with lysosomes, and they uncoat to disgorge their contents into the cytoplasm. Subsequent work in our labs, in the Walensky lab, at Aileron Therapeutics and in other labs has shown the active transport of stapled peptides to be remarkably general. Peptides having a wide variety of amino acid sequences, staple locations within the sequence and types of staple – $i,i+4$ and $i,i+7$ – utilize endocytic vesicle trafficking to gain entry into cells. Of course, not all of these are imported with equal efficiency, and in particular, basic stapled peptides seem to be imported preferentially, while those having net negative charge are often taken up less efficiently (Bernal et al., 2006). For this reason, and also to promote aqueous solubility, it has become common practice in labs doing stapled peptide research to replace non-essential acidic residues with neural or basic ones, and also to replace non-essential hydrophobic residues with polar or basic residues.

SAHB$_A$ was found to induce apoptotic cell death in Jurkat cells and a panel of human leukemia cells at 2–10 mM concentrations, whereas the single-point mutant negative control peptide and the unmodified Bid BH3 peptide were completely inactive over this concentration range (Walensky et al., 2004). To assess the activity of SAHB$_A$ in mouse tumor models, we teamed up with Andrew Kung, who engineered refractory RS4;1:1 human leukemia cells to express luciferase, and then transplanted these into γ-irradiated severe combined immunodeficient mice. The leukemia was allowed to engraft for 3 days, and then either SAHB$_A$ or the inactive SAHB$_A$ point mutant was administered by tail vein injection at 10 mg/kg/day for 3 days, and the tumor burden was assessed by whole body luminescence imaging using the Xenogen IVIS system (Caliper Life Sciences, Hopkinton, MA). A separate cohort of animals was treated with the Dimethylsulfoxide (DMSO) vehicle and imaged on the same schedule. The SAHB$_A$-treated animals showed a highly significant reduction in tumor burden and mean survival time as compared with the vehicle-treated animals, while the animals treated with the point mutant peptide fared little better than the vehicle-treated control (Walensky et al., 2004). These data provided compelling evidence that stapled peptides can exert a potent anti-tumor effect in a mouse model of cancer.

The pharmaceutical optimization of SAHB$_A$ and related stapled BH3 domain peptides is ongoing, an effort being undertaken by Aileron Therapeutics. Among the numerous contributions made to stapled peptide development by scientists at Aileron is the discovery that relatively routine sequence optimization of these peptides can yield versions with half-lives of 12–24 hours in rats, with high-capacity but low-affinity binding to serum albumin, and having an *in vivo* clearance mechanism primarily consisting of hepatobiliary excretion rather than renal filtration, with no evidence of processing by cytochrome P450 oxidases (T. Sawyer and R. Kapeller, pers. comm.). Taken together, these pharmacological data paint a picture for stapled peptides that is completely distinct from the norm for peptides, one that bodes well for the emergence of this as an entirely new class of therapeutics for treatment of human disease.

If small molecules are applicable to ~10% of all human targets and biologicals are applicable to another 10%, what fraction of targets currently considered "undruggable" will become tractable with the advent of stapled peptides as drugs? The answer to this question, of course, remains to be determined, but it is not beyond the realm of possibility that stapled peptide technology could ultimately render another 10% of human targets druggable. Thus, the need to discover additional new classes of molecules beyond stapled peptides continues to remain an urgent one.

ACKNOWLEDGMENTS

I am deeply grateful to the donors of the High-Tech Fund of the DFCI for funding our research on stapled peptides. An immense debt of gratitude is due Stan Korsmeyer for his friendship, indefatigable optimism and seminal contributions to the translational development of stapled peptides. The promise of stapled peptides might have gone unrealized absent the tremendous advances made by Loren Walensky, and I cannot express in words how extraordinarily talented and capable this young scientist is in whom I am so proud and to whom I am so indebted. I also thank Chris Schafmeister for his key role in the design of the peptide stapling system, and to Fed Bernal for his lovely work on targeting hDm2 and hDmx. Thanks are also due our collaborators on the SAHB$_A$ project, namely Andrew Kung and Gerhard Wagner. Finally, I thank the present members of the Verdine Lab, especially the peptide stapling team, for their ongoing efforts.

REFERENCES

Bartoli, S., Fincham, C.I., and Fattori, D. 2007. Fragment-based drug design: combining philosophy with technology. *Curr Opin Drug Discov Devel* **10**:422–429.

Bernal, F., Tyler, A.F., Korsmeyer, S.J., Walensky, L.D., and Verdine, G.L. 2006. Reactivation of the p53 tumor suppressor pathway by a stapled p53 peptide. *J Am Chem Soc* **129**:2456–2457.

Binz, H.K., Amstutz, P., and Pluckthun, A. 2005. Engineering novel binding proteins from nonimmunoglobulin domains. *Nat Biotechnol* **23**: 1257–1268.

Bird, G.H., Bernal, F., Pitter, K., and Walensky, L.D. 2008. Synthesis and biophysical characterization of stabilized alpha-helices of BCL-2 domains. *Methods Enzymol* **446**:369–386.

Blackwell, H.E. et al. 2001. Ring-closing metathesis of olefinic peptides: design, synthesis, and structural characterization of macrocyclic helical peptides. *J Org Chem* **66**:5291–5302.

Boal, A.K. et al. 2007. Facile and e-selective intramolecular ring-closing metathesis reactions in 310-helical peptides: a 3D structural study. *J Am Chem Soc* **129**:6986–6987.

Chao, D.T. and Korsmeyer, S.J. 1998. BCL-2 family: regulators of cell death. *Annu Rev Immunol* **16**:395–419.

Danial, N.N. 2007. BCL-2 family proteins: critical checkpoints of apoptotic cell death. *Clin Cancer Res* **13**:7254–7263.

Danial, N.N. and Korsmeyer, S.J. 2004. Cell death: critical control points. *Cell* **116**:205–219.

Danial, N.N. et al. 2008. Dual role of proapoptotic BAD in insulin secretion and beta cell survival. *Nat Med* **14**:144–153.

Gavathiotis, E. et al. 2008. BAX activation is initiated at a novel interaction site. *Nature* **455**:1076–1081.

Gross, A., McDonnell, J.M., and Korsmeyer, S.J. 1999. BCL-2 family members and the mitochondria in apoptosis. *Genes Dev* **13**:1899–1911.

Grubbs, R.H. 2004. Olefin metathesis. *Tetrahedron* **60**:7114–7140.

Hajduk, P.J. and Greer, J. 2007. A decade of fragment-based drug design: strategic advances and lessons learned. *Nat Rev Drug Discov* **6**:211–219.

Hanahan, D. and Weinberg, R.A. 2000. The hallmarks of cancer. *Cell* **100**: 57–70.

Henchey, L.K., Jochim, A.L., and Arora, P.S. 2008. Contemporary strategies for the stabilization of peptides in the alpha-helical conformation. *Curr Opin Chem Biol* **12**:692–697.

Hershberger, S.J., Lee, S.G., and Chmielewski, J. 2007. Scaffolds for blocking protein-protein interactions. *Curr Top Med Chem* **7**:928–942.

Hopkins, A.L. and Groom, C.R. 2002. The druggable genome. *Nat Rev Drug Discov* **1**:727–730.

Karle, I.L. and Balaram, P. 1990. Structural characteristics of alpha-helical peptide molecules containing Aib residues. *Biochemistry* **29**:6747–6756.

Kim, Y.W. and Verdine, G.L. 2009. Stereochemical effects of all-hydrocarbon tethers in i,i+4 stapled peptides. *Bioorg Med Chem Lett* **19**:2533–2536.

Korsmeyer, S.J. 1992. Bcl-2 initiates a new category of oncogenes: regulators of cell death. *Blood* **80**:879–886.

Marshall, G.R. et al. 1990. Factors governing helical preference of peptides containing multiple alpha,alpha-dialkyl amino acids. *Proc Natl Acad Sci U S A* **87**:487–491.

Miller, S.J., Blackwell, H.E., and Grubbs, R.H. 1996. Application of ring-closing metathesis to the synthesis of rigidified amino acids and peptides. *J Am Chem Soc* **118**:9606–9614.

Nuttall, S.D. and Walsh, R.B. 2008. Display scaffolds: protein engineering for novel therapeutics. *Curr Opin Pharmacol* **8**:609–615.

Petros, A.M., Olejniczak, E.T., and Fesik, S.W. 2004. Structural biology of the Bcl-2 family of proteins. *Biochim Biophys Acta* **1644**:83–94.

Pitter, K., Bernal, F., Labelle, J., and Walensky, L.D. 2008. Dissection of the BCL-2 family signaling network with stabilized alpha-helices of BCL-2 domains. *Methods Enzymol* **446**:387–408.

Russ, A.P. and Lampel, S. 2005. The druggable genome: an update. *Drug Discov Today* **10**:1607–1610.

Sattler, M. et al. 1997. Structure of Bcl-xL-Bak peptide complex: recognition between regulators of apoptosis. *Science* **275**:983–986.

Schafmeister, C.J., Po, J., and Verdine, G.L. 2000. An all-hydrocarbon cross-linking system for enhancing the helicity and metabolic stability of peptides. *J Am Chem Soc* **122**:5891–5892.

Skerra, A. 2007. Alternative non-antibody scaffolds for molecular recognition. *Curr Opin Biotechnol* **18**:295–304.

Stumpp, M.T., Binz, H.K., and Amstutz, P. 2008. DARPins: A new generation of protein therapeutics. *Drug Discov Today* **13**:695–701.

Verdine, G.L. and Walensky, L.D. 2007. The challenge of drugging undruggable targets in cancer: lessons learned from targeting BCL-2 family members. *Clin Cancer Res* **13**:7264–7270.

Walensky, L.D. 2006. BCL-2 in the crosshairs: tipping the balance of life and death. *Cell Death Differ* **13**:1339–1350.

Walensky, L.D. et al. 2004. Activation of apoptosis in vivo by a hydrocarbon-stapled BH3 helix. *Science* **305**:1466–1470.

Walensky, L.D. et al. 2006. A stapled Bid BH3 helix directly binds and activates BAX. *Mol Cell* **24**:199–210.

BASAL BODIES: THEIR ROLES IN GENERATING ASYMMETRY

SUSAN K. DUTCHER

Department of Genetics, Washington University School of Medicine, St. Louis, Missouri

I. BASAL BODIES AND CENTRIOLES

Microtubule organizing centers play an important role in all eukaryotic cells. The structure and composition of these organelles varies. In an animal cell, the centrosome is the focal point for the cytoplasmic microtubules and is often near the nucleus of the cell. The animal centrosome consists of the pericentriolar material that includes gamma (γ)-tubulin and its associated γ-tubulin ring complex and the pair of centrioles. Most fungi and diatoms nucleate their microtubules from a spindle pole body that is embedded in the nuclear envelope. Flowering plants have microtubule organizing sites that are ill defined morphologically but are distributed around the nuclear envelope.

The role of centrioles in the animal centrosome has been unclear. In the 1890s, Boveri suggested that the structures were the "soul of the cell," while more recently centrioles have been relegated to consideration as a vestigial organelle. The connection between centrioles and their related organelle, basal bodies, suggests an important and essential role in ciliary and flagellar assembly, and more recently, a role in cellular signaling. Electron microscopic studies in the 1950s and 1960s showed that centrioles and basal bodies share a common structure. They are both cylinders of triplet microtubules; centrioles are near the nucleus during interphase and at the poles of the spindle in mitosis, while basal bodies are anchored to the plasma membrane at the base of the cilia or flagella. To understand the role of the centriole and its related organelle, the basal body, this laboratory have used the unicellular, green alga *Chlamydomonas reinhardtii* for genetic, biochemical, and computational experiments. In particular, this report will briefly describe the structure of basal bodies and

17

The Harvey Lectures, Series 102, pages 17–50
©2010 by John Wiley & Sons, Inc.

then turn to their functions in the cell. In particular, the role of the unique nonequivalence between the two basal bodies or centrioles that arises via their mode of duplication is explored.

II. BASAL BODY REPLICATION AND GENERATION OF BASAL BODY NONEQUIVALENCE

Basal bodies and centrioles usually duplicate once per cell. In mammalian cells, duplication occurs in conjunction with the S phase. It is followed by the separation of the centrioles at the beginning of the G2/M phase. Each pole of the spindle contains one new or daughter centriole and one old or mother centriole (Kochanski and Borisy, 1990). Unlike in mammalian cells, duplication of a new basal body begins during mitosis, and intermediates in the duplication known as probasal bodies in *Chlamydomonas* are present in early G1 and remain arrested during the next cell cycle. These probasal bodies then elongate to form two full-length basal bodies. The timing of this elongation event is likely to be in mitosis, but could be in S/G2 as these events occur very rapidly in *Chlamydomonas*. Several groups have observed that elongation is preceded by the loss of the distal striated fiber to allow the old basal bodies to lie parallel to one another rather than lying perpendicularly to one another (Gould, 1975).

Electron microscopy has been used to follow the events of basal body initiation and elongation. The formation of the probasal body begins with an annulus of amorphous material within 1 hour of cytokinesis (Gould, 1975); this is likely to be the same ring of amorphous material observed in mature basal bodies (O'Toole et al., 2003). Johnson and Porter (1968) reported that the first event of basal body duplication is the formation of a ring of nine singlet microtubules. The B and C microtubules are added in an irregular manner. However, Gould (1975) never observed singlet microtubules. At present, these events are not well documented.

Centrioles in animal cells are distinguishable from each other in G1 of the cell cycle. The older or mother centriole is marked by the presence of subdistal fibers at the distal end and has the ability to assemble a primary cilium (Vorobjev and Chentsov, 1982). Ninein, ODF2, CEP110, dynactin, and epsilon (ε)-tubulin are associated with the mother centriole, and immunoelectron microscopy has shown that ninein, CEP110, and ε-tubulin localize to the subdistal fibers. The acquisition of these epitopes and the ability to assemble a cilium is termed maturation. In each cell cycle, the daughter centriole will undergo maturation in G2 to become a mature centriole and acquire competence to assemble a cilium in the next cell cycle. All basal bodies in the respiratory ciliated epithelium are associated with ODF2 (Ishikawa et al., 2005; Cao et al., 2006). Thus, basal bodies in a ciliated epithelium must undergo maturation in a non-cell cycle mediated manner. In *Chlamydomonas*, the younger basal body must undergo maturation in the G1 phase of the cell cycle so that it can assemble a flagellum. The two mature basal bodies in *Chlamydomonas* are associated with ε-tubulin.

This difference in the age of basal bodies or centrioles generates nonequivalence between the two basal bodies or centrioles. The nonequivalence of the basal bodes will be discussed with respect to the phenotype in various mutants, and the generation of ciliary asymmetry, cellular asymmetry, and mitotic asymmetries.

III. Structure of Basal Bodies

Many features of *Chlamydomonas* basal bodies are shared with these structures in most organisms. These characteristics include the presence of nine triplet microtubules, which consist of a 13-protofilament tubule, called the A tubule, and two partial tubules with 11 protofilaments, each called the B and C tubule; together, these three tubules are referred to as the "microtubule blade." A second attribute is the orthogonal position of the two basal bodies. In an interphase *Chlamydomonas* cell, the two basal bodies are found at right angles to one another at the anterior end of the cell (Ringo, 1967) (Fig. 2.2a). This arrangement is found in most centriole pairs, and in *Chlamydomonas* it may allow the two flagella to beat effectively in opposite directions. In association with the two mature basal bodies are two probasal bodies. In contrast to the 200-nm length of the mature basal body, probasal bodies have triplet microtubules that are only 40-nm long. At the proximal end of both basal bodies and probasal bodies is a ring of amorphous material that is ~20 nm in depth (Fig. 2.1b-1,c-1). Distal to this region is the pinwheel, which in cross-sectional images, appears as a hub with a series of radiating spokes from the center of the basal body to the triplet microtubules where they connect to the A tubule by triangular projections. This

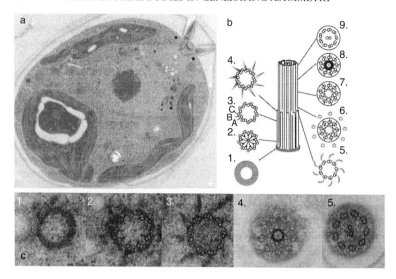

Fig. 2.1. (a) Electron micrograph of a *Chlamydomonas* cell preserved by cryofixation methods to illustrate the position of the pair of basal bodies (indicated by *) in the cell at the anterior end. (Image supplied by Dr Eileen O'Toole, University of Colorado.) (b) Diagram of basal body and transition zone structures. The distance from the base to the tip is approximately 450 nm. (1.) Ring of amorphous material. (2.) Triplet microtubules, also known as microtubule blades, with pinwheel. (3.) The A, B, and C tubules of one triplet microtubule are indicated. The central shaft is devoid of the pinwheel, but γ-tubulin has been localized to this region. (4.) The transition fibers are shaded and begin at the distal end of the basal body. They extend and remain attached through the doublet microtubules of the transition zone. (5.) Transition fibers at the proximal end of the transition zone. (6.) Pores of the ciliary necklace; they are the termination of the transition fibers. They surround the transition zone and hold the basal body onto the membrane. The stellate fibers of the proximal transition zone have two distinct regions: (7.) the proximal stellate fibers; and (8.) the distal stellate fiber system with osmiophilic ring and small inner fiber. (9.) Proximal end of flagellum prior to appearance of axonemal substructures. (From O'Toole et al., 2003.) (c.) Electron micrographic images. (1.) Corresponds to diagram 1 of panel (b). (2.) Corresponds to diagram 2 of panel (b). (3.) Corresponds to diagram 4 of panel (b). (4.) Corresponds to diagram 8 of panel (b). (5.) Corresponds to diagram 9 of panel (b). (From Dutcher, 2001.)

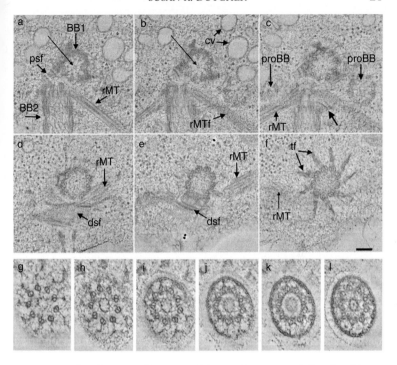

Fig. 2.2.   Selected tomographic images of the transition zone and fibers in wild-type *Chlamydomonas*. (a) One basal body is shown in cross section (BB1) and the other in longitudinal view (BB2). The proximal end of the basal body consists of an amorphous electron-dense ring, and there is ninefold symmetrical pinwheel in the center (arrow). A two-membered rootlet microtubule is seen in the lower right (rMT). (b) The pinwheel of BB1 has a center formed from three rings (arrow). Contractile vacuoles are present (cv). (c) Two probasal bodies (proBB) are near the mature basal bodies. Fibers connected the basal body to the rootlet microtubule (arrow). (d and e) The distal striated fiber and rMTs are indicated. (f) Transitional fibers radiate out from the triplets at the distal end of the basal body proper. (g) The proximal region at the end of the transition fibers (between diagrams 5 and 6 in Fig. 2.1b) has doublet microtubules and Y-shaped connecters (arrows). (h and i) The first stellate fiber array consists of a nine-pointed star with a central hub formed from electron-dense triangular points. The distal tips of the transition fibers and Y-shaped connecters form knobs or pores that attach to the plasma membrane (*). (j) The first stellate array is replaced by a central, amorphous disk (arrow). (k) The second stellate array at the distal end of the transition zone consists of a nine-pointed star with an elaborate center. (l) The two stellate fiber arrays in longitudinal section look like an osmiophilic H. (With permission from *Molecular Biology of the Cell*; from O'Toole et al., 2003.)

pinwheel structure is found in most basal bodies and often in the probasal bodies. Along the spokes of the pinwheel are bulges with unknown function (O'Toole et al., 2003). The pinwheel has a depth of about 40 nm. In longitudinal sections, the pinwheel can be seen as a series of plates or tiers (Cavalier-Smith, 1974).

The cylinder of the triplet microtubules extends for an additional 200 nm and is referred to as the basal body shaft or basal body proper. Immunoelectron microscopy has suggested that γ-tubulin may be present in the interior of the shaft (Silflow et al., 2001). At the end of the basal body proper is the transition zone in which the C tubule of the triplet microtubules ends and doublet microtubules begin.

IV. The Transition Zone

The distal part of the structure is known as the transition zone; it is the region that transitions from triplet to doublet microtubules. It is associated with two fibrous structures; they are the transition fibers and the stellate fibers. The transition fibers are striated and triangular at their attachment to the basal bodies (Ringo, 1967; O'Toole et al., 2003). They bend and form a cage around the distal regions of the basal body that attach the basal bodies to the plasma membrane (Fig. 2.1c-3). The ends of the fibers appear similar to the Y-shaped connectors seen in *Spermatozopsis similis* (Lechtreck et al., 1999) (Fig. 2.2g). Lastly, there are knob- or pore-like structures termed "the ciliary necklace," which serve as anchor points for the basal body (Weiss et al., 1977). These are found at the ends of the transition fibers. Y-shaped fibers have been described in a wide range of ciliated cells, including those from metazoans. The protein P210 was identified in the wall-less alga *S. similis*, and antibodies to it localize to the Y-shaped connectors (Lechtreck et al., 1999). The protein is also present in nascent basal bodies and basal body precursors (Lechtreck and Grunow, 1999; Lechtreck and Bornens, 2001). The antibody to *S. similis* P210 similarly stains the transition zone of some *Chlamydomonas* species (Schoppmeier and Lechtreck, 2002). The transition zone is likely to be the homolog of the subdistal fibers found on the mother centriole in animal cells.

The interior of the microtubule blade of the transition zone contains two distinct stellate fiber systems (Fig. 2.1b-7,8 and 2.1c-4, Fig. 2.2h–l).

Each system has nine points around a central ring, and the nine points attach to the A tubule. The proximal stellate fiber has osmiophilic staining in the center; the points and the ring show similar staining. The distal stellate fibers lack the center staining, and the ring stains much more intensely than the points (Ringo, 1967; O'Toole et al., 2003). A 10-nm disk of amorphous material separates the two stellate fiber systems (O'Toole et al., 2003). Stellate fibers are unique to the green algae and are not found in all basal bodies. The stellate fibers form the osmiophilic H-shaped structure observed in longitudinal sections (see Fig. 2.5).

V. Distal Striated Fibers and Proximal Fibers

Centrin is a 20-kD EF-hand protein that forms calcium-sensitive, contractile fibers that are found in at least three distinct locations in the cell (Salisbury et al., 1988). In interphase cells, it is present in the distal striated fiber that holds the older and younger basal bodies to each other. The distal striated fiber attaches to microtubule blades 9, 1, and 2 of the basal bodies (Hoops and Witman, 1983) and appear as bundles of oriented fibers. The distal striated fiber is connected to the basal bodies by an additional fiber system (O'Toole et al., 2003; E. O'Toole and S.K. Dutcher, unpublished observations). Centrin is also present in another set of fibers that attach at microtubule blades 7 and 8 and connect the basal bodies to the nucleus; these fibers are termed the nucleo-basal body connectors. They also are referred to as the rhizoplast. Centrin is also present in the transition zone in the stellate fibers and in the inner dynein arms of the flagellar axoneme. Several lines of evidence suggest that centrin fibers contract in vivo with changes in calcium concentration (Salisbury et al., 1987).

During interphase, the mature basal bodies are connected to each other by proximal striated fibers as well (Fig. 2.3). The composition of the proximal fiber is unknown. The mature basal body and the probasal body are connected to each other by a 6-nm fiber (Gould, 1975), but again, its composition is unknown. In mammalian cells, the attachment of centrioles at the proximal end is mediated by a protein called rootletin (Bahe et al., 2005). It is a coiled-coil protein and is regulated by Nek2, Never-in-mitosis A (NimA) like kinase. Phosphorylation of it allows for the separation of the centriole pair in mammalian cells.

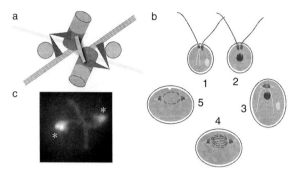

Fig. 2.3. The fiber systems attached to the *Chlamydomonas* basal body complex. (a) The mature basal bodies are shown in red, the transition zones in peach, and the probasal bodies in pink. The microtubular rootlets have four microtubules (orange) or two microtubules (yellow) and attach at specific triplet microtubules of the basal body. The distal (solid) and proximal (striped) striated fibers are shown in light blue (here, the latter are partially hidden by the former). They connect the two mature basal bodies at the two ends. The lateral fibers are shown in green. They connect the mature basal body to its daughter probasal body across the rootlet microtubules. (b) Changes in the fiber systems during the cell cycle. (1) During interphase, the basal bodies and transition zones are continuous with the flagella. The rootlet microtubules are adjacent to the plasma membrane. One of the four-membered rootlet microtubules lies adjacent to the eyespot (rose). (2) Another view of interphase cells illustrates that the basal bodies are connected to the nucleus (dark blue) and to each other by centrin fibers (light blue). (3) At preprophase, the flagella are lost. The probasal bodies elongate. The distal and proximal striated fibers are lost. (4) The two-membered rootlet microtubules shorten. The centrioles (without transition zones) are found at the poles of the spindle. The four-membered rootlet microtubules arc over the spindle. The eyespot is disassembled. (5) Cytokinesis is initiated at one end of the cell. This will be followed by extension of the two-membered rootlet microtubules and reassembly of the striated fibers, as well as assembly of new rootlet microtubules and of a new eyespot in association with one of the new four-membered microtubular rootlets. (Image from Dutcher, 2003.) (c) The rootlet microtubules arching across the spindle as visualized by antibody to acetylated α-tubulin. The poles of the spindle are visualized by antibody to centrin (*). (Image from Jessica Esparza, Washington University.)

VI. Mutations with Altered Basal Bodies

A. *BLD2 and BLD10 loci*

The *bld2-1* allele was isolated in a screen for cells that are unable to mate. Because flagella are required for the initial stages of partner recognition, numerous aflagellate strains were identified with this approach. The *bld2* strain was shown to have aberrant basal bodies (Goodenough and St. Clair, 1975). Instead of assembling triplet microtubules, most cells had singlet microtubules and the cylinder was often only 40–50 nm in length instead of 200 nm. The *BLD2* locus encodes ε-tubulin, a tubulin isoform that is found only in ciliated organisms with triplet microtubules (Chang and Stearns, 2000; Dutcher et al., 2002; Dutcher, 2003). The mutant allele has a stop codon at amino acid 9. ε-Tubulin localizes to the transition fibers in *Chlamydomonas* (Dutcher et al., 2002) and to the sub-distal fibers of mammalian centrioles (Chang et al., 2003). It does not appear to be a component of the microtubule cylinder. Yeast two-hybrid experiments with mouse ε-tubulin failed to identify another tubulin partner (S.K. Dutcher and L. Li, unpublished observations). The mechanism by which ε-tubulin stabilizes the triplet microtubules remains unclear.

The *bld10-1* strain, like *bld2-1*, is an aflagellate strain that does not assemble microtubule blades, but it is unclear if the amorphous material is present (Matsuura et al., 2004). This mutant represents the most severe basal body assembly defect in *Chlamydomonas* to date. BLD10 is a 170-kD coiled-coil protein that has an ortholog in humans. Immunoelectron microscopy suggests that the BLD10 protein localizes to the pinwheel in the proximal end of the basal body. Like *bld2* mutants, *bld10-1* has disorganized rootlet microtubules and centrin fibers. Unlike *bld2*, disorganized spindles are observed in this strain (Matsuura et al., 2004). Truncation of the last third of the C-terminus of the BLD10 protein results in basal bodies with eightfold rather than ninefold symmetry. The spokes of the pinwheel are reduced in length (Hiraki et al., 2007).

The role of basal bodies in meiosis in *Chlamydomonas* has remained controversial. Cavalier-Smith (1974) observed the loss of transition fibers and the transition zone soon after the formation of zygotes, and reported that basal bodies are lost in the meiotic cells. Alternatively, it was reported that basal bodies are visible at prophase of the first meiotic division (Triemer and Brown, 1976). The *bld2-1* and *bld10-1* mutations have a recessive meiotic defect, which suggests strongly that basal bodies play a role in meiosis (Preble et al., 2001).

VII. The Role of Basal Body Nonequivalence

Just as the two centrioles in an animal cell are not equivalent, the two basal bodies in *Chlamydomonas* are not equivalent. One is a parental or

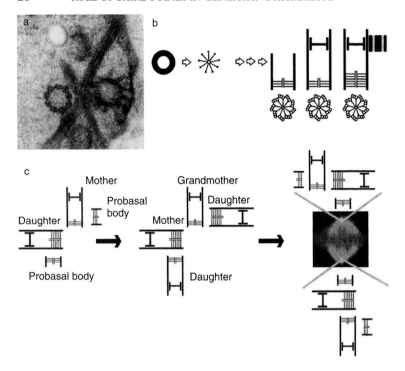

Fig. 2.4.   Replication and maturation of basal bodies. (a) Two mature basal bodies lying at right angle to each other and two probasal bodies at 12 and 6 o'clock. They are separated by rootlet microtubules. (b) Pathway of assembly and maturation in *Chlamydomonas*. An amorphous ring is postulated to form first and then the nine spokes of the pinwheel. The microtubule blades are added (not shown). The pinwheel is postulated to change with the age of the basal body, and the distal striated fiber is added to attach the two mature basal bodies to each other. (c) The basal bodies are associated in a stereotypical fashion. One basal body in each pair is always a daughter; the other is a mother, a grandmother, a great-grandmother, and/or older. In *Chlamydomonas*, the probasal bodies are present throughout the cell cycle rather than just before elongation, as is observed in animal cells.

mother basal body that is at least one generation older, and the other is the daughter basal body that was formed in the previous cell cycle (Fig. 2.4) as confirmed by direct observations of dividing cells (Gaffel, 1988; Holmes and Dutcher, 1989). In animal cells, the older centriole is distinguished by the presence of subdistal fibers and often the loss of the pinwheel at the base, while in *Chlamydomonas*, there are no obvious changes in the morphology of the transition fibers between the mother and the daughter basal body. However, the *Chlamydomonas* basal body appears to have alterations in the depth of the pinwheel. Beech et al. (1991) observed that the number of pinwheel tiers was related to the cell cycle age of the basal bodies, with the older basal body having fewer tiers in other algae. The number of tiers in *Chlamydomonas* basal bodies varies from two to seven tiers, with three and four tiers being most common (Geimer and Melkonian, 2004). This would suggest that an older basal body generally has two or three (46% of the basal bodies) and the younger basal bodies have four to seven tiers (54% of the basal bodies).

A. Nonequivalence of Basal Bodies Revealed by Mutants

The existence of several mutant strains that are able to assemble only one of the two flagella reveals that the two basal bodies are not equivalent and the loss of these proteins have different consequences for basal bodies of differing ages.

The *uni1* mutants. The five independently isolated *uni1* strains (*uni1-1* to *uni1-5)* have wild-type morphologies in the basal body proper but have differences in their transition zones. The older of the two basal bodies assembles additional stellate fibers, while the younger basal body fails to assemble any stellate fibers (Fig. 2.5). As the name of the mutant implies, the *uni⁻* cells have only one flagellum. The older basal body assembles a flagellum and the younger basal body gains the ability to assemble a flagellum after one cycle by direct observation (Holmes and Dutcher, 1989). In addition, the younger basal body fails to assemble any transition fibers (McVittie, 1972; Huang et al., 1982). Although the gene product of the *UNI1* locus is not known at the present time, it is likely to be important to the conversion and maturation of the daughter in the first cell cycle into a mother basal body.

This mutant raises the interesting question of whether the maturation is time dependent or whether it is cell cycle dependent. The older basal body has been in the cytoplasm longer and may just have more to complete a process that takes longer in the *uni1* mutant strain. *Chlamydomonas* biology offers two ways to address this question. First, the cell cycle time can be slowed by using minimal medium or lower light intensities. Neither of these alternations changes the fraction of cells with one

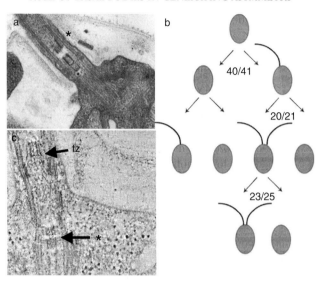

Fig. 2.5.   Basal body nonequivalence in *uni1* and *uni3* mutants. (a) Electron micrograph of the basal bodies from a *uni1* cell. One basal body has additional material (*) in the transition zone, while the other basal body has no transition zone at all. (b) Pedigree results for dividing *uni3* cells. (c) Electron micrograph of the basal bodies from a *uni3* cell. Note that the osmiophilic material is present both in the transition zone (tz) and in the basal body proper (*).

flagellum. Second, the timing of flagellar assembly was analyzed in light/dark-synchronized cells where the time between rounds of cell divisions was more than 20 hours. In each division period, most cells in these conditions divide twice or three times in rapid succession to produce four or eight cells, with time between divisions being only 1 hour. When there are four cells produced, two of the cells have parental basal bodies that are only 2 hours old, one has a basal body that is nearly 24 hours old, and the fourth basal body is potentially ancient, but at least 48 hours old. Nevertheless, the four *uni1* sister cells quartet reveals that all four cells assemble a single flagellum in approximate synchrony (Holmes and Dutcher, 1992). The *uni1* cells have two basal bodies that differ in age by only 1 hour behave identically to cells with a pair of basal bodies that differ in age by more than 20 hours. It appears that the Uni⁻ phenotype of this mutant results form a difference in cell cycle age rather than a difference in their chronological age.

The uni3 mutant. The *UNI3* gene also reveals nonequivalence between basal bodies that extends over not one cell cycle but multiple cell cycles. The *uni3-1* allele is a deletion that removes another tubulin isoform, delta (δ)-tubulin, which is also found only in organisms with triplet microtubules (Dutcher and Trabuco, 1998; Chang and Stearns, 2000; Garreau de Loubresse et al., 2001). Unlike the Uni1 mutant flagellar phenotype, the Uni3 mutant phenotype is heterogeneous. One-half of the cells are aflagellate, about one-quarter of the cells have a single flagellum, and the remaining cells have two flagella. The cells with two flagella can have flagella of different lengths (Dutcher and Trabuco, 1998; Garreau de Loubresse et al., 2001).

Electron microscopic tomography revealed several phenotypes of the *uni3* basal bodies. First, the basal bodies have doublet rather than triplet microtubules along most of the length of the basal body. However, there is a short stretch containing triplet microtubules just before the beginning of the transition zone (Dutcher and Trabuco, 1998; O'Toole et al., 2003). Second, the distal striated fiber fails to assemble. This suggests a role for the C tubule in the attachment of the distal striated fiber. Third, there are additional stellate fibers formed in the cylinder shaft more proximally than observed in wild-type cells. This is a phenotype that differs from the Uni1 phenotype where the extra fibers are found at the distal end in the transition zone. These extra fibers are in the cylinder of the basal body proper and there are often multiple stellate fibers formed. This suggests that the presence of doublet microtubules may be the signal for the assembly of the stellate fiber system (O'Toole et al., 2003).

To ask about the generation of the heterogeneous flagellar assembly phenotype in the *uni3* mutant, pedigree experiments were performed. One hypothesis suggests that the phenotype could be generated stochastically, which would predict that the progeny of any one division would reflect the probabilities in the population. For example, the chance of two uniflagellate sister cells would be $1/4 \times 1/4$ or $1/16$, or the chance of an aflagellate and a uniflagellate sister cell would be $2 \times (1/2 \times 1/4)$ or $1/4$. The phenotype of the two sister cells would be independent on the phenotype of the parent. Alternatively, a second hypothesis suggests that the phenotype of the progeny would be dependent on the phenotype of the parent and perhaps the age of the basal bodies and could be predicted by the phenotype of the parent. Using growing conditions that result in a single round of cell division, it was found that the two sisters could be predicted by the phenotype of the parental cell (Fig. 2.5). Aflagellate cells produced one aflagellate cell and one uniflagellated cell (40 out of 41 cells). Uniflagellated cells produced one aflagellate cell and one biflagellated cell (20 out of 21 cells). Biflagellated cells produced one aflagellate cell and one biflagellated cell (23 out of 25 cells) (Dutcher and Trabuco, 1998) (Fig. 2.5). In the absence of δ-tubulin, the nonequivalence of the basal bodies is obvious for multiple generations.

To explain the pedigree, we suggest that the aflagellate cell has a mother and a daughter basal body and neither is able to recruit the proteins needed for flagellar assembly. The uniflagellate cell is postulated to have a grandmother and a daughter basal body. The grandmother basal body has gained the ability to recruit proteins needed for flagellar assembly. The biflagellate cell is postulated to have a great-grandmother basal body and a daughter basal body. The great-grandmother is postulated to be able to recruit proteins that not only promote flagellar assembly on the great-grandmother basal body but also promote assembly on the daughter basal body. This recruitment of proteins in a cell with the great-grand-mother does not survive cell division. When the biflagellate cells divide, it gives rise to a biflagellate cell and an aflagellate cell again. We suggest that the daughter basal body does not retain the ability to recruit proteins and build a flagellum. Thus, the pedigree experiments suggest that the phenotype of the parental strains can predict the number of flagella that are assembled in the daughter cells based on a model of basal body age.

The uni2 mutant. The *UNI2* gene is also needed for maturation and mutants show nonequivalence. The populations of cells have zero, one, and two flagella in about the same ratios as observed in the *uni3* mutant. As observed in both *uni1* and *uni3* cells, the stellate fiber material in the transition zone is abnormal; however, it is more similar to the phenotype observed in the *uni1* cell. The UNI2 protein is a homolog of a mammalian centriole protein called CEP120 (Andersen et al., 2003; Piasecki et al., 2008). The UNI2 protein exists in a phosphorylated and unphosphorylated form; the phosphorylated form accumulates at the end of mitosis with the assembly of the transition zone and new flagella (Piasecki et al., 2008).

From the analysis of these mutants, it is clear that progress through the cell cycle can "age" a basal body. To identify proteins that are needed for aging the basal bodies, we constructed double mutants with *uni3*. Both *UNI1* and *UNI2* are needed for the aging of the *uni3* basal bodies. The double mutants assemble no flagella. Thus, CEP120, UNI1, and δ-tubulin are needed to properly age the basal bodies (Dutcher and Trabuco, 1998). The other gene that may be needed for aging is *TTL6* that is required for glutamylation of tubulin (Wloga et al., 2008). The two basal bodies differ in their level of the posttranslational modification of added extra glutamates onto the carboxy terminus of tubulin (Lechtreck and Geimer, 2000).

B. *Cellular Asymmetry That Arises from Basal Body Nonequivalence*

When one observes a *Chlamydomonas* cell, there are several asymmetric features in the cell. One of these is the eyespot that plays a role in phototaxis (Melkonian and Robenek, 1984), and the other is the position of the structures needed for zygote fusion in the two gametic mating-types. Placement of the eyespot and the mating structures is likely to be dependent on a set of specialized microtubules called the rootlet microtubules. Melkonian (1978) showed that the pattern of the rootlet microtubules is a valuable tool for the classification of algae. Microtubular rootlets are composed of four bundles of microtubules that originate from the basal body region and extend under the plasma membrane toward the equator of the cell. In *Chlamydomonas*, there are two bundles with two microtubules and two bundles with four microtubules (Figs. 2.3a and 2.4a). By electron microscopy, these microtubules are tightly opposed to one another compared with the cytoplasmic microtubules; they are arranged in a pattern of three microtubules in a row over a single microtubule. Most green algae show the 4-X-4-X pattern, where X is the number of microtubules in the other rootlet. By light microscopy, rootlet microtubules can be distinguished from cytoplasmic microtubules by their stability (Holmes and Dutcher, 1989) and by the presence of acetylated α-tubulin, which is detected by an antibody to this posttranslational modification on lysine 40 of α-tubulin (Piperno and Fuller, 1985; LeDizet and Piperno, 1986; Holmes and Dutcher, 1989).

Rootlet microtubules are found associated with specific triplet microtubules of the basal body. The four-membered microtubular rootlets terminate near triplets 2 and 3, and the two-membered microtubular rootlets are near triplet 9; together they form a distinct cross-shaped structure (Fig. 2.4a). The rootlet microtubules associate with the cleavage furrow during mitosis. Unlike the cytoplasmic microtubules, these microtubules do not disassemble at the transition from interphase to mitosis but remain associated with the duplicated pairs of basal bodies and arch over the spindle, curving at the midpoint (Fig. 2.3c).

The rootlet microtubules play a role in positioning other organelles. The four-membered microtubular rootlet from the daughter basal body ends in the vicinity of the eyespot, which is in the chloroplast. The two-membered microtubular rootlet from the daughter basal body is associated with contractile vacuoles and with the mating structure (Weiss, 1984). The eyespot in wild-type cells is always positioned near the daughter basal body (Holmes and Dutcher, 1989). As a result of the asymmetric, but different, placement of mating structures in the *plus* and *minus* mating-type cells, fusion of the gametes always results in dikaryons with parallel pairs of flagella and the eyespots on the same side of the newly mated cell (Fig. 2.6). In addition, the basal bodies, as they enter mitosis, must retain their same orientation. The two mature basal bodies are attached at specific microtubule blades. As mentioned above, they must break the attachment between mature basal bodies and form attachments with the new daughter basal bodies. This event requires that the basal bodies rotate so that the microtubules that faced the old basal body now face the daughter basal body. Since the basal bodies can be marked by the presence of a contractile vacuole, their behavior can be monitored by the behavior of the contractile vacuole. Both show a clockwise orientation. Furthermore, the spindle pole also retains a fixed orientation. The older basal body at each pole remains closer to the cleavage furrow, and the younger one will be near the site for the new eyespot. These asymmetries are likely due to the asymmetries of the basal body pair (Holmes and Dutcher, 1989).

C. Ciliary Asymmetry that Arises from the Nonequivalence of the Basal Bodies

Mutants defective in phototaxis. Motile cilia move cells (*Chlamydomonas*, protists, and sperm as examples) or move fluids along an epithelial surface (the respiratory tract or the oviduct as examples). The movement is generated by the activity of dynein arms, which are large multiprotein

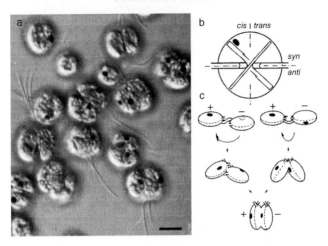

Fig. 2.6. Asymmetry of mating in *Chlamydomonas* may be related to basal body maturity. (a) A field of 15-minute-old dikaryons taken with differential interference contrast microscopy. Relative to the position of the eyespot (dark ovals), the mating occurs along the *syn* side of one gamete and the *anti* side of the other gamete (see panel b.). The outcome of mating is that the flagella from the mature basal body (*trans*) are oriented together and the flagella from the daughter basal body (*cis*) are oriented together. This arrangement allows the cells to continue to orient toward a light source. The scale bar is 5 μm. (b) *syn* is defined as the half of the cell with the eyespot, and *anti* is defined as the half of the cell lacking the eyespot. The mating-type plus structure is formed on the *anti* side. (c) During mating, flagella can adhere in a *cis*-to-*cis* position (right) or in a *cis*-to-*trans* position (left). In either case, the cells fuse with the eyespots on the same side. In the diagram, the left gamete is mating-type plus and the right gamete is mating-type minus. (Reprinted with permission from *Journal of Cell Science*; from Holmes and Dutcher, 1989.)

complexes that walk along microtubules (Porter and Sale, 2000). The two flagella of *Chlamydomonas* have different properties from each other, and this may be important for the responses to environmental signals that include light and chemicals that allow the cells to orient. These differences include differences in the beat frequency and differences in response to calcium ions. Most biflagellate algae do not have two identical flagella, and the differences correlate with the parental/daughter history of the basal bodies (Beech et al., 1991). The older basal body in the cell is termed the *trans* basal body, and the younger basal body is termed the *cis* basal body after their position with respect to the eyespot. In demembranated cell models, low calcium concentrations inactivate the *trans* flagellum, while at higher concentrations (10^{-6} to 10^{-7} M Ca^{+2}), the *cis* flagellum is inactive (Kamiya and Witman, 1984). In living cells that have only a single flagellum, the *trans* flagellum has a 30% higher beat frequency than the *cis* flagellum. When both are present, they beat at the frequency of the *cis* flagellum (Omoto and Brokaw, 1985; Kamiya and Hasegawa, 1987; Rüffer and Nultsch, 1987). Further analysis suggested that changes in the waveform and not the beat frequency of the *cis* and *trans* flagella regulate the cell during phototactic turning (Rüffer and Nultsch, 1987). During phototactic turning, the *trans* flagellum has a decreased front amplitude and is less effective. Mutants that are defective in phototaxis provide an approach to understand what generates the differences between the flagella templated by the old and new basal bodies and how the differences change the activity of the flagella. To date, mutants that act to alter the activity of the flagella have been identified in the I1 dynein.

One class of mutants that have ineffective phototaxis affects the I1 or *f* dynein (King et al., 1986). Mutants that affect the outer dynein arms or other inner dynein arm genes do not perturb phototaxis. The I1 dynein arm is a two-headed dynein that is composed of two dynein heavy chains (1α and 1β), three intermediate chains (IC140, IC138, and IC98), and five light chains (TcTex1, TcTex2b, LC8, LC7a, LC7b) and is found every 96 nm along the inner circumference of the A tubule in the flagellar axoneme (reviewed by Wirschell et al., 2007). The I1 arm is important for the control of microtubule sliding and the shape of the flagellar waveform (Brokaw and Kamiya, 1987). Control of the flagellar waveform is likely to be mediated through the phosphorylation and dephosphorylation of the I1 intermediate chain IC138. All of the mutants identified to date that fail to assemble the I1 arm show phototaxis defects (Myster et al., 1997; Mastronarde et al., 1992; Porter et al., 1992, 1999; Perrone et al., 2003). In addition, two mutants (*mia1*, *mia2*), which were isolated as phototaxis defective, assemble the I1 dynein arm and show hyperphosphorylation of IC138 (King and Dutcher, 1997). Phosphorylation of IC138 control sliding of the microtubules in vitro (Habermacher and Sale, 1997).

We postulated that when the eyespot receives illumination, a signal is transduced to the flagellar membrane where Ca^{+2} ion channels open. Elevated Ca^{+2} concentrations would increase the activity of the phosphatase relative to the kinase in the *trans* flagellum, causing the dephosphorylation of the IC138-kD intermediate chain and the subsequent activation of the I1 dynein complex. The *trans* flagellum would then beat with an increased front amplitude relative to the *cis* flagellum, and the cell would turn toward the light source. Because the cell rotates, the eyespot becomes shaded and the flagellar Ca^{+2} ion channels would be closed. The reduction in Ca^{+2} ion concentration would increase the activity of the kinase relative to the phosphatase. IC138 become phosphorylated and the I1 dynein complex is activated. The *trans* flagellum would then beat with a decreased front amplitude relative to the *cis* flagellum, and the cell would continue to turn toward the light source. The loss of the entire I1 complex in the *pf9/ida1*, *ida2*, and *ida3* strains results in an inability to perform phototaxis because an end effector of the phototaxis signal transduction pathway is absent. Without the presence of the *f* dynein complex, the cells were unable to alter direction in response to light stimuli. In the *mia⁻* strains, the amount of altered phosphorylated species of the 138-kD substrate was increased. This increase would be predicted to result in a less effective waveform and a concurrent loss in the response to phototactic signals. The *mia⁻* mutations could shift the balance of kinase/phosphatase activity in favor of the kinase regardless of transduction signals. The phosphatase, kinase inhibitor, and substrate are just three examples of gene products whose alteration or loss of activity could result in the skewing the phosphorylation regulatory balance. Alternatively, the mutations could affect the asymmetric localization of regulatory components between the *cis* and *trans* flagella. Previously, we postulated that the levels of phosphorylated IC138 were different between the *cis* and the *trans* flagellum.

Mutants in phototactic directionality. Wild-type cells generally swim toward light of low intensity. A variant isolated from the environment shows an aberrant behavior (Smyth and Ebersold, 1985). The mutant called *agg1* is not defective in phototaxis as the cells sense light, but responds incorrectly by swimming away from the light source rather than toward it. The use of RNA interference techniques to reduce the expression level of two genes, now called *AGG2* and *AGG3*, reveals that these transformants swim away from a light source that attracts wild-type cells (Iomini et al., 2006). These two genes provide some insights into the role of nonequivalence in determining the directionality of movement toward light.

The AGG2 and AGG3 proteins were identified by a biochemical approach to find proteins in isolated flagellar membranes from

Chlamydomonas that are detergent resistant (DRMs). DRMs are membrane fragments with proteins that are enriched in sterols and sphingolipids but deprived of unsaturated glycerophospholipids. These fractions from isolated *Chlamydomonas* membranes and separated on sucrose gradients contained four bands by polyacrylamide gel electrophoresis and were identified by mass spectroscopy. Bands 1 and 3 encode paralogs with molecular weights of 19,570 and 15,624 daltons. They show 63% identity and 68% similarity to each other, and both have transmembrane domains. The AGG2 gene encodes the smaller protein. *AGG3* encodes a protein in band 3 that has a PLAC8 or DUF614 domain that is broadly conserved in eukaryotes, but has an unknown function. It has a molecular weight of 21,414 daltons. It also has a paralog with a molecular weight of 21,386 daltons. These two proteins differ by only seven amino acids, and both contain a 156-amino-acid predicted flavodoxin domain (Iomini et al., 2006). The antibody to *AGG3* is likely to recognize its paralog and localize to the flagellar matrix.

The Agg⁻ pathway(s) must function in wild-type cells to permit positive phototaxis. Due to the helical nature of the swimming, cells receive alternating periods of light intensity if they are not swimming perpendicularly to a light source. Wild-type *Chlamydomonas* shows positive phototaxis because one of the two flagella responds preferentially to a light signal. As described above, the *cis* flagellum is located in the hemisphere with the eyespot and the *trans* flagellum is on the other side. In a genetically wild-type strain, the *trans* flagellum is responsive under most physiological conditions. When the eyespot transmits reception of the light signal, the *trans* flagellum responds with a larger amplitude of its waveform (indicated by the larger shaded area in Fig. 2.7) than the *cis* flagellum, and the cell turns toward the light. As the cell rotates, the eyespot becomes shaded and the *cis* flagellum has the larger amplitude to reinforce the turning toward the light (Rüffer and Nultsch, 1991; King and Dutcher, 1997). This process results in the continuous turning of the cell toward the light. The response of the *trans* flagellum to a light signal has been referred to as flagellar dominance. Analysis of wild-type and *agg1* strains by high-speed cinematography show that the flagellar dominance is reversed in the *agg1* strain. While in *agg1* cells, the *cis* flagellum has the larger amplitude when lit and the *trans* flagellum has the larger amplitude when shaded. Since the process is symmetrically reversed, the *agg1* cells respond to light by swimming away from it (Rüffer and Nultsch, 1998). We

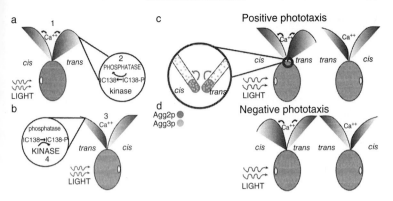

Fig. 2.7. (a) Schematic model of a hypothesis for asymmetric phosphoregulation of I1 dynein activity and its role in phototaxis. When the eyespot (yellow) is lighted (a), the rhodopsin signal is transduced to the flagellar membrane where Ca^{+2} ion channels open (1). Elevated Ca^{+2} concentrations would increase the activity of the phosphatase relative to the kinase in the *trans* flagellum (2). This would result in the dephosphorylation of IC138 and subsequent activation of the I1 dynein complex. The *trans* flagellum would then beat with an increased front amplitude (shaded area under flagella) relative to the *cis* flagellum, and the cell would turn toward the light source. As the cell rotates, the eyespot becomes shaded (b) and the flagellar Ca^{+2} ion channels would close (3). The resultant reduction in Ca^{+2} ion concentration would increase the activity of the kinase relative to the phosphatase (4). This would lead to phosphorylation of IC138 and the deactivation of the I1 dynein complex. The *trans* flagellum would then beat with a decreased front amplitude relative to the *cis* flagellum, and the cell would continue to turn toward the light source. (c) A schematic model for a possible role of AGG2 and AGG3 in orientation during phototaxis. Positive phototaxis – as in panel (a) above, the *trans* flagellum in positive phototaxis is activated as indicated by the larger shaded area under the flagellum. The signals mediated by AGG2 and AGG3 (red and orange dots) are hypothesized to inhibit the dynein arms in the *cis* flagellum (purple line) and activate the dynein arms in the *trans* flagellum (blue line). As the cell turns away from the light during along its helical swimming path, the signal is lost and the two flagella have more similar waveforms with a slightly larger amplitude for the *cis* flagellum. (d) During negative phototaxis, the cell shows the reversed response. The *cis* flagellum responds because the inhibition is lost and the *trans* flagellum is not activated.

suggest that the AGG1, AGG2, and AGG3 proteins block transmission of the light signal in the *cis* flagellum and reinforce the signal in the *trans* flagellum. When these proteins are reduced, the *cis* flagellum responds as the dominant flagellum.

AGG2 localizes to the proximal region of flagella that still has 9 + 0 microtubules, and is likely to extend into the region with 9 + 2 microtubules. Several axonemal proteins have a discrete localization in the flagella that could function as anchoring sites for AGG2. Nephrocystin-1, which is mutated in type I nephronophthisis, localizes to the proximal region of motile cilia of the respiratory epithelium (Schermer et al., 2005) as do several NIMA-related kinases (Mahjoub et al., 2004, 2005). To test if FA2, a NIMA-related kinase in *Chlamydomonas*, anchors AGG2, localization of AGG2 was examined in *fa2-4* cells. Its localization remains indistinguishable in *fa2-4* cells compared with wild-type cells.

It is interesting to remember that *AGG2* and *AGG3* both have paralogs. *AGG3* has two paralogs (DQ408775 and FAP 191) that were found by mass spectroscopy of isolated flagella and were enriched in the membrane-matrix fraction in additional studies (Pazour et al., 2005). A fourth paralog (155892) was not found in these studies and may not be a flagellar protein. Both the *AGG2* and *AGG3* paralogs are likely to have arisen by a recent duplication as the paralogs are tightly linked to each other. It is interesting to speculate that different pairs of paralogs may be found in the *cis* and *trans* flagella to help impart different responses to light signals. Tagged versions of the paralogs will be needed to pursue this hypothesis. In addition, it is not clear that the RNAi experiments knockout down only one or the other paralog.

VIII. Templating Ciliary Asymmetries by the Basal Body

One major function of basal bodies is the templating of flagella. The analysis of the *bld2-2 rgn1-1* strain provides evidence for this function. In this double mutant, basal bodies with singlet, doublet, and triplet microtubules as well as missing microtubules are observed by electron microscopy (Preble et al., 2001). If the basal bodies are missing a microtubule blade, then the axoneme is missing a doublet microtubule (Fig. 2.8). Basal bodies with a singlet microtubule will assemble a flagellar axoneme with a singlet microtubule. The capacity for direct templating of the axonemal microtubules resides in the basal body.

Several of the genes needed to assemble the I1 inner dynein arm were first identified in a suppressor screen with the *pf10* mutant (Dutcher et al., 1988). The *pf10* mutant is required for wild-type motility (Lewin, 1954; McVittie, 1972; Randall and Starling, 1972) and shows a nearly symmetrical waveform instead of the asymmetrical one observed in wild-

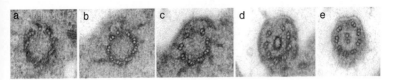

Fig. 2.8. The basal body plays a role in templating the flagellar axoneme. Serial section electron micrographs of the basal body and flagellar axoneme. (a and b) The basal body proper has a singlet rather than a triplet microtubule. (c and d) The transition zone is missing one of the blades. (e) The flagellar axoneme is missing a blade.

type cells. It shows a unique phenotype for a motility mutant. In newly mated dikaryons formed between most paralyzed flagellar (*pf*) mutant strains and a wild-type strain, paralysis will be relieved by the incorporation of the wild-type gene product into the pair of flagella from the *pf* parent (Luck et al., 1977). The *pf10* mutant is not rescued by the presence of wild-type cytoplasm. If the flagella are removed and allowed to reassemble in the mixed cytoplasm, two of the flagella show normal motility and two of them show the altered motility. Thus, the rescue of this phenotype is either hampered by the lack of a pool of wild-type proteins or by a necessity for a point in the cell cycle other than G1 (Dutcher, 1986). Similar behaviors were observed for *fla9-1*, a temperature-sensitive flagellar assembly mutant (Adams et al., 1982; Dutcher, 1986), for *uni1-1*, and for *uni3-1*. This failure to rescue in dikaryons is consistent with a basal body defect (Dutcher, 1986). Isolated basal bodies from the *pf10* mutant are missing two polypeptides with molecular weights of 60,000 daltons (Fig. 2.9).

Suppressors of the *pf10* motility phenotype identified seven extragenic loci. They are *LIS1*, *LIS2*, *BOP1*, *BOP2*, *BOP3*, *BOP4*, and *BOP5* (Dutcher et al., 1988). The *BOP2* locus encodes IC140 (Perrone et al., 1998) and *BOP5* encodes IC138 (Hendrickson et al., 2004) of the I1 inner dynein arm. Thus, the *PF10* locus is of great interest in that it appears to affect the assembly of basal bodies based on its lack of two polypeptides, and it is likely to affect the assembly of the I1 dynein inner arm based on the suppressors that have been identified. PF10 may be important for assembling and specifying potential differences between the *cis* and *trans* inner dynein arm I1.

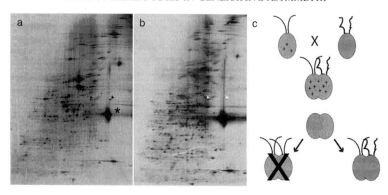

Fig. 2.9.   (a) Two-dimensional gel electrophoresis of isolated basal bodies from wild-type *Chlamydomonas*. Apparent molecular weight is on the vertical-axis, and the apparent pI is on the horizontal-axis. α- and β-tubulins are indicted by asterisk (*) and are estimated to make up 65% of the protein by mass. The two proteins that are missing or diminished in *pf10* preparations are marked by black arrowheads. (b). Two-dimensional gel of isolated basal bodies from *pf10* cells. White arrowheads indicate the two polypeptides missing or reduced from these preparations compared with preparations from wild-type cells. (c) The outcome of dikaryons using wild-type cells and *pf10* cells. Most flagellar mutants give the result on the left of four motile flagella, while *pf10* matings produce two motile and two aberrant motility flagella (on the right).

IX. Mitotic Asymmetry

Mutations that affect the assembly of basal body microtubules cause profound defects in the orientation and placement of the rootlet microtubules. These include mutations in *BLD2*, *UNI3*, and *BLD10* (Ehler et al., 1995; Dutcher and Trabuco, 1998; Preble et al., 2001; Matsuura et al., 2004). In these mutants, the cross-shaped pattern of rootlet microtubules is lost and the bundles of microtubules begin to fray as revealed by an increased number of structures that are identified by antibodies to acetylated α-tubulin (Ehler et al., 1995). These mutants show defects in the placement of the cleavage furrow with respect to the nucleus (Ehler et al., 1995; Dutcher and Trabuco, 1998; Preble et al., 2001; Matsuura et al., 2004). However, these frayed rootlet microtubules remain associated with the cleavage furrow.

These mutants share another phenotype. They are hypersensitive to the action of the microtubule depolymerization drug Taxol and arrest in mitosis. Given that several proteins that depolymerize the spindle micro-tubules can influence the stability the mitotic microtubule, the ability to localize the protein katanin p80 was investigated. Several lines of evidence suggest that the transition fibers on the basal bodies are the site of docking for intraflagellar transport (IFT) motors and proteins. Immunofluores-cence microscopy with antibodies to motors as well as IFT proteins shows accumulation around this region. Second, immunoelectron microscopy with antibodies to IFT52 shows localization to the transition fibers (Deane et al., 2001). Finally, the presence of transition fibers is correlated with the ability to assemble a flagellum. For example, the *uni1-1* mutant has one basal body with transition fibers and a flagellum, and one basal body without transition fibers and without a flagellum (Huang et al., 1982). Similarly, in the pair of mammalian centrioles, only the centriole with subdistal fibers, which appear to be orthologous to the transition fibers of *Chlamydomonas*, can assemble a primary cilium (Rieder and Borisy, 1982; Vorobjev and Chentsov, 1982; Chretien et al., 1997).

Basal bodies from the *bld2* and *bld10* mutant strains are unable to localize katanin p80 (J. Esparza, L. Li, A. Albee, and S.K. Dutcher, unpublished observations). The *uni3* mutant shows a heterogeneous localization pattern. There are cells that lack proper katanin80 localiza-tion; there are cells with reduced amounts, and cell with increased amounts. It is likely that these correspond to cells with varying ages of their basal bodies. Further work will be needed to ask if other proteins also require intact basal bodies for their localization.

X. Why Are Basal Body and Centriole Asymmetries Important?

We suggest that the aging of basal bodies and centrioles may have multiple functions. The aging of centrioles and their role in the mitotic spindle placement may be a key function. Elegant work in *Drosophila melanogaster* suggest the older centriole is generally associated with the stem cell in the asymmetric division that occurs to generate a stem cell and a differentiating cell in the niche of the male germ line (Yamashita et al., 2007; Yamashita and Fuller, 2008). Adult stem cells divide asymmetrically, and in this example, the mother centriole remains anchored near the niche, while the daughter centriole migrates to the opposite side of the cell so that it will be localized to the differentiat-ing cell.

A second function may reside in the signaling role of the cilia. It may be important to allow the cilium templated by the older centriole/basal body to commence signaling before the cilium templated by the younger centriole/basal body (Anderson and Stearns, 2007). Primary cilia are responsible for the signaling generated by a large number of pathways. It has become clear that many signaling molecules are localized to the primary cilium. Among these are proteins in the hedgehog pathway, the platelet derived growth factor (PDGF) receptor, the polycystins, and melanocortin 4 receptor 1 (Eggenschwiler and Anderson, 2007; Berbari et al., 2008).

XI. Genes with Connections to Human Disease

Since many flagellar and basal body proteins show high levels of similarity with proteins in other ciliated organisms but no similarity to proteins in nonciliated organisms, it is possible to use BLAST searches to find homologs among ciliated organisms that are missing from nonciliated organisms (Avidor-Reiss et al., 2004; Li et al., 2004). A comparison between *Chlamydomonas* and human identifies over 4,000 homologs. About 700 of these are missing from *Arabidopsis*, a seed plant that lacks basal bodies and cilia. Only 300 of these are reciprocal best matches between *Chlamydomonas* and human. This data set was used to identify the human disease gene Bardet Biedl syndrome 5 gene (*BBS5*) on human chromosome 2 (Li et al., 2004).

Bardet Biedl syndrome (BBS) is a rare autosomal recessive human disease that was first recognized in the late 1800s. It is associated with renal disease, similar to nephronophthisis, central obesity, polydactyly, retinal degeneration, anosmia, hypogenitalia, and hypertrophy of the heart. At present, there are 13 genes that have been identified in human BBS patients. *BBS6* is the only one without a clear homolog in *Chlamydomonas*. *BBS5* was identified using the conservation of ciliary genes from humans to *Chlamydomonas*, and the absence of these genes in nonciliated organisms such as *Arabidopsis* or yeast. The *BBS5* gene was mapped in a family from Newfoundland. The region of interest was 14 Mb and contained 230 genes. Only two of the 230 genes were present in the comparative cilia database. One of these genes was unknown and contained a splice site mutation in the affected members of the family. Affected members of three additional families from Kuwait carry mutations that result in premature stop codons in this gene (Li et al., 2004). Antibodies raised to the mouse protein showed localization around the basal bodies in mouse ependymal cells. Nachury and coworkers have demonstrated that the BBS protein form a complex and are needed for transport of six-pass membrane proteins into cilia (Nachury et al., 2007; Berbari et al., 2008; Nachury, 2008).

Other proteins found by comparative genomics are likely to play roles in cilia. These include Tubby superfamily protein (TUSP), fibrocystin, MKS1, UNC119, which is also known as HRP4, and seahorse and qilin, which were found to cause embryonic lethality and cystic kidney disease in zebrafish (Li et al., 2004).

XII. Conclusions

Ideas about basal bodies and cilia have changed dramatically over the last few years, from being considered irrelevant to be key players in many pathways. With the rediscovery that cilia are found on many cells in humans and play important roles in sensing the environment, it is important to discover more about their assembly and their functions; in particular, the role of basal bodies in generating asymmetry in cells and organisms. The asymmetry of the basal bodies does not arise during their formation but through their maturation or aging. The older basal bodies provide a site not only for assembly but also for the recruitment of additional proteins to aid in its role in ciliary assembly and in the mitotic spindle.

Acknowledgments

I want to thank the members of my lab over the last 25 years for their contributions and enthusiasm. In particular, the work described here owes a large debt of gratitude to Jeff Holmes, Bill Inwood, Steve King, Mary Porter, and Eileen O'Toole. I also want to thank my family for their support and love. I thank the National Institutes of Health and the Searle Scholars for financial support.

References

Adams, G.M., Huang, B., and Luck, D.J. 1982. Temperature-sensitive, assembly-defective flagella mutants of Chlamydomonas reinhardtii. *Genetics* **100**: 579–586.

Andersen, J.S., Wilkinson, C.J., Mayor, T., Mortensen, P., Nigg, E.A., and Mann, M. 2003. Proteomic characterization of the human centrosome by protein correlation profiling. *Nature* **426**:570–574.

Anderson, C.T. and Stearns, T. 2007. The primary cilium: what once did nothing, now does everything. *J Musculoskelet Neuronal Interact* **7**:299.

Avidor-Reiss, T., Maer, A.M., Koundakjian, E., Polyanovsky, A., Keil, T., Subramaniam, S., and Zuker, C.S. 2004. Decoding cilia function: defining specialized genes required for compartmentalized cilia biogenesis. *Cell* **117**:527–539.

Bahe, S., Stierhof, Y.D., Wilkinson, C.J., Leiss, F., and Nigg, E.A. 2005. Rootletin forms centriole-associated filaments and functions in centrosome cohesion. *J Cell Biol* **171**:27–33.

Beech, P.L., Heimann, K., and Melkonian, M. 1991. Development of the flagellar apparatus during the cell cycle of unicellular alga. *Protoplasma* **164**:23–27.

Berbari, N.F., Lewis, J.S., Bishop, G.A., Askwith, C.C., and Mykytyn, K. 2008. Bardet-Biedl syndrome proteins are required for the localization of G protein-coupled receptors to primary cilia. *Proc Natl Acad Sci U S A* **105**: 4242–4246.

Brokaw, C.J. and Kamiya, R. 1987. Bending patterns of Chlamydomonas flagella: IV. Mutants with defects in inner and outer dynein arms indicate differences in dynein arm function. *Cell Motil Cytoskeleton* **8**:68–75.

Cao, W., Gerton, G.L., and Moss, S.B. 2006. Proteomic profiling of accessory structures from the mouse sperm flagellum. *Mol Cell Proteomics* **5**:801–810.

Cavalier-Smith, T. 1974. Basal body and flagellar development during the vegetative cell cycle and the sexual cycle of Chlamydomonas reinhardii. *J Cell Sci* **16**:529–556.

Chang, P. and Stearns, T. 2000. Delta-tubulin and epsilon-tubulin: two new human centrosomal tubulins reveal new aspects of centrosome structure and function. *Nat Cell Biol* **2**:30–35.

Chang, P., Giddings, T.H. Jr., Winey, M., and Stearns, T. 2003. Epsilon-tubulin is required for centriole duplication and microtubule organization. *Nat Cell Biol* **5**:71–76.

Chretien, D., Buendia, B., Fuller, S.D., and Karsenti, E. 1997. Reconstruction of the centrosome cycle from cryoelectron micrographs. *J Struct Biol* **120**: 117–133.

Deane, J.A., Cole, D.G., Seeley, E.S., Diener, D.R., and Rosenbaum, J.L. 2001. Localization of intraflagellar transport protein IFT52 identifies basal body transitional fibers as the docking site for IFT particles. *Curr Biol* **11**: 1586–1590.

Dutcher, S.K. 1986. Genetic properties of linkage group XIX in Chlamydomonas reinhardtii. *Basic Life Sci* **40**:303–325.

Dutcher, S.K. 2001. The tubulin fraternity: alpha to eta. *Curr Opin Cell Biol* **13**:49–54.

Dutcher, S.K. 2003. Elucidation of basal body and centriole functions in Chlamydomonas reinhardtii. *Traffic* **4**:443–451.

Dutcher, S.K. 2003. Long-lost relatives reappear: identification of new members of the tubulin superfamily. *Curr Opin Microbiol* **6**:634–640.

Dutcher, S.K. and Trabuco, E.C. 1998. The UNI3 gene is required for assembly of basal bodies of Chlamydomonas and encodes delta-tubulin, a new member of the tubulin superfamily. *Mol Biol Cell* **9**:1293–1308.

Dutcher, S.K., Gibbons, W., and Inwood, W.B. 1988. A genetic analysis of suppressors of the PF10 mutation in Chlamydomonas reinhardtii. *Genetics* **120**:965–976.

Dutcher, S.K., Morrissette, N.S., Preble, A.M., Rackley, C., and Stanga, J. 2002. Epsilon-tubulin is an essential component of the centriole. *Mol Biol Cell* **13**:3859–3869.

Eggenschwiler, J.T. and Anderson, K.V. 2007. Cilia and developmental signaling. *Annu Rev Cell Dev Biol* **23**:345–373.

Ehler, L.L., Holmes, J.A., and Dutcher, S.K. 1995. Loss of spatial control of the mitotic spindle apparatus in a Chlamydomonas reinhardtii mutant strain lacking basal bodies. *Genetics* **141**:945–960.

Gaffel, G.P. 1988. The basal body-root complex of Chlamydomonas reinhardtii during mitosis. *Protoplasma* **143**:118–129.

Garreau de Loubresse, N., Ruiz, F., Beisson, J., and Klotz, C. 2001. Role of delta-tubulin and the C-tubule in assembly of Paramecium basal bodies. *BMC Cell Biol* **2**:4.

Geimer, S. and Melkonian, M. 2004. The ultrastructure of the Chlamydomonas reinhardtii basal apparatus: identification of an early marker of radial asymmetry inherent in the basal body. *J Cell Sci* **117**:2663–2674.

Goodenough, U.W. and St. Clair, H.S. 1975. BALD-2: a mutation affecting the formation of doublet and triplet sets of microtubules in Chlamydomonas reinhardtii. *J Cell Biol* **66**:480–491.

Gould, R.R. 1975. The basal bodies of Chlamydomonas reinhardtii. Formation from probasal bodies, isolation, and partial characterization. *J Cell Biol* **65**:65–74.

Habermacher, G. and Sale, W.S. 1997. Regulation of flagellar dynein by phosphorylation of a 138-kD inner arm dynein intermediate chain. *J Cell Biol* **136**:167–176.

Hendrickson, T.W., Perrone, C.A., Griffin, P., Wuichet, K., Mueller, J., Yang, P., Porter, M.E., and Sale, W.S. 2004. IC138 is a WD-repeat dynein intermediate chain required for light chain assembly and regulation of flagellar bending. *Mol Biol Cell* **15**:5431–5442.

Hiraki, M., Nakazawa, Y., Kamiya, R., and Hirono, M. 2007. Bld10p constitutes the cartwheel-spoke tip and stabilizes the 9-fold symmetry of the centriole. *Curr Biol* **17**:1778–1783.

Holmes, J.A. and Dutcher, S.K. 1989. Cellular asymmetry in Chlamydomonas reinhardtii. *J Cell Sci* **94**:273–285.

Holmes, J.A. and Dutcher, S.K. 1992. Genetic approaches to the study of cytoskeletal structure and function in Chlamydomonas. In *The Cytoskelton of the Algae* (ed. D. Menzel). CRC Press; Boca Raton, LA, pp. 347–368.

Hoops, H.J. and Witman, G.B. 1983. Outer doublet heterogeneity reveals structural polarity related to beat direction in Chlamydomonas flagella. *J Cell Biol* **97**:902–908.

Huang, B., Ramanis, Z., Dutcher, S.K., and Luck, D.J. 1982. Uniflagellar mutants of Chlamydomonas: evidence for the role of basal bodies in transmission of positional information. *Cell* **29**:745–753.

Iomini, C., Li, L., Mo, W., Dutcher, S.K., and Piperno, G. 2006. Two flagellar genes, AGG2 and AGG3, mediate orientation to light in Chlamydomonas. *Curr Biol* **16**:1147–1153.

Ishikawa, H., Kubo, A., Tsukita, S., and Tsukita, S. 2005. Odf2-deficient mother centrioles lack distal/subdistal appendages and the ability to generate primary cilia. *Nat Cell Biol* **7**:517–524.

Johnson, U.G., and Porter, K.P. 1968. Fine structure of cell division in *Chlamydomonas reinhardi*. Basal bodies and microtubules. *J Cell Biol* **38**: 406–425.

Kamiya, R. and Witman, G.B. 1984. Submicromolar levels of calcium control the balance of beating between the two flagella in demembranated models of Chlamydomonas. *J Cell Biol* **98**:97–107.

Kamiya, R. and Hasegawa, E. 1987. Intrinsic difference in beat frequency between the two flagella of Chlamydomonas reinhardtii. *Exp Cell Res* **173**: 299–304.

King, S.J. and Dutcher, S.K. 1997. Phosphoregulation of an inner dynein arm complex in Chlamydomonas reinhardtii is altered in phototactic mutant strains. *J Cell Biol* **136**:177–191.

King, S.M., Otter, T., and Witman, G.B. 1986. Purification and characterization of Chlamydomonas flagellar dyneins. *Methods Enzymol* **134**:291–306.

Kochanski, R.S. and Borisy, G.G. 1990. Mode of centriole duplication and distribution. *J Cell Biol* **110**:1599–1605.

Lechtreck, K.F. and Bornens, M. 2001. Basal body replication in green algae – when and where does it start? *Eur J Cell Biol* **80**:631–641.

Lechtreck, K.F. and Geimer, S. 2000. Distribution of polyglutamylated tubulin in the flagellar apparatus of green flagellates. *Cell Motil Cytoskeleton* **47**:219–235.

Lechtreck, K.F. and Grunow, A. 1999. Evidence for a direct role of nascent basal bodies during spindle pole initiation in the green alga Spermatozopsis similis. *Protist* **150**:163–181.

Lechtreck, K.F., Teltenkotter, A., and Grunow, A. 1999. A 210 kDa protein is located in a membrane-microtubule linker at the distal end of mature and nascent basal bodies. *J Cell Sci* **112**(Pt. 11):1633–1644.

LeDizet, M. and Piperno, G. 1986. Cytoplasmic microtubules containing acetylated alpha-tubulin in Chlamydomonas reinhardtii: spatial arrangement and properties. *J Cell Biol* **103**:13–22.

Lewin, R.A. 1954. Mutants of Chlamydomonas moewusii with impaired motility. *J Gen Microbiol* **11**:358–363.

Li, J.B., Gerdes, J.M., Haycraft, C.J., Fan, Y., Teslovich, T.M., May-Simera, H., Li, H., Blacque, O.E., Li, L., Leitch, C.C., Lewis, R.A., Green, J.S., Parfrey, P.S., Leroux, M.R., Davidson, W.S., Beales, P.L., Guay-Woodford, L.M., Yoder, B.K., Stormo, G.D., Katsanis, N., and Dutcher, S.K. 2004. Comparative genomics identifies a flagellar and basal body proteome that includes the BBS5 human disease gene. *Cell* **117**:541–552.

Luck, D., Piperno, G., Ramanis, Z., and Huang, B. 1977. Flagellar mutants of Chlamydomonas: studies of radial spoke-defective strains by dikaryon and revertant analysis. *Proc Natl Acad Sci U S A* **74**:3456–3460.

Mahjoub, M.R., Qasim Rasi, M., and Quarmby, L.M. 2004. A NIMA-related kinase, Fa2p, localizes to a novel site in the proximal cilia of Chlamydomonas and mouse kidney cells. *Mol Biol Cell* **15**:5172–5186.

Mahjoub, M.R., Trapp, M.L., and Quarmby, L.M. 2005. NIMA-related kinases defective in murine models of polycystic kidney diseases localize to primary cilia and centrosomes. *J Am Soc Nephrol* **16**:3485–3489.

Mastronarde, D.N., O'Toole, E.T., McDonald, K.L., McIntosh, J.R., and Porter, M.E. 1992. Arrangement of inner dynein arms in wild-type and mutant flagella of Chlamydomonas. *J Cell Biol* **118**:1145–1162.

Matsuura, K., Lefebvre, P.A., Kamiya, R., and Hirono, M. 2004. Bld10p, a novel protein essential for basal body assembly in Chlamydomonas: localization to the cartwheel, the first ninefold symmetrical structure appearing during assembly. *J Cell Biol* **165**:663–671.

McVittie, A. 1972. Flagellum mutants of Chlamydomonas reinhardii. *J Gen Microbiol* **71**:525–540.

Melkonian, M. 1978. Structure and significance of cruciate flagellar root systems in green algae: comparative investigations in species of *Chlorosarcinopsis*. *Plant Syst. Evol.* **130**:265–292.

Melkonian, M. and Robenek, H. 1984. The eyespot apparatus of flagellated green algae: a critical review. *Prog Phycol Res* **3**:193–268.

Myster, S.H., Knott, J.A., O'Toole, E., and Porter, M.E. 1997. The Chlamydomonas Dhc1 gene encodes a dynein heavy chain subunit required for assembly of the I1 inner arm complex. *Mol Biol Cell* **8**:607–620.

Nachury, M.V. 2008. Tandem affinity purification of the BBSome, a critical regulator of Rab8 in ciliogenesis. *Methods Enzymol* **439**:501–513.

Nachury, M.V., Loktev, A.V., Zhang, Q., Westlake, C.J., Peranen, J., Merdes, A., Slusarski, D.C., Scheller, R.H., Bazan, J.F., Sheffield, V.C., and Jackson, P.K. 2007. A core complex of BBS proteins cooperates with the GTPase Rab8 to promote ciliary membrane biogenesis. *Cell* **129**: 1201–1213.

O'Toole, E.T., Giddings, T.H., McIntosh, J.R., and Dutcher, S.K. 2003. Three-dimensional organization of basal bodies from wild-type and delta-tubulin deletion strains of Chlamydomonas reinhardtii. *Mol Biol Cell* **14**: 2999–3012.

Omoto, C.K. and Brokaw, C.J. 1985. Bending patterns of Chlamydomonas flagella: II. Calcium effects on reactivated Chlamydomonas flagella. *Cell Motil* **5**:53–60.

Pazour, G.J., Agrin, N., Leszyk, J., and Witman, G.B. 2005. Proteomic analysis of a eukaryotic cilium. *J Cell Biol* **170**:103–113.

Perrone, C.A., Yang, P., O'Toole, E., Sale, W.S., and Porter, M.E. 1998. The Chlamydomonas IDA7 locus encodes a 140-kDa dynein intermediate chain required to assemble the I1 inner arm complex. *Mol Biol Cell* **9**:3351–3365.

Perrone, C.A., Tritschler, D., Taulman, P., Bower, R., Yoder, B.K., and Porter, M.E. 2003. A novel dynein light intermediate chain colocalizes with the retrograde motor for intraflagellar transport at sites of axoneme assembly in Chlamydomonas and mammalian cells. *Mol Biol Cell* **14**: 2041–2056.

Piasecki, B.P., Lavoie, M., Tam, L.W., Lefebvre, P.A., and Silflow, C.D. 2008. The Uni2 phosphoprotein is a cell cycle regulated component of the basal body maturation pathway in Chlamydomonas reinhardtii. *Mol Biol Cell* **19**:262–273.

Piperno, G. and Fuller, M.T. 1985. Monoclonal antibodies specific for an acetylated form of alpha-tubulin recognize the antigen in cilia and flagella from a variety of organisms. *J Cell Biol* **101**:2085–2094.

Porter, M.E. and Sale, W.S. 2000. The 9 + 2 axoneme anchors multiple inner arm dyneins and a network of kinases and phosphatases that control motility. *J Cell Biol* **151**:F37–F42.

Porter, M.E., Power, J., and Dutcher, S.K. 1992. Extragenic suppressors of paralyzed flagellar mutations in Chlamydomonas reinhardtii identify loci that alter the inner dynein arms. *J Cell Biol* **118**:1163–1176.

Porter, M.E., Bower, R., Knott, J.A., Byrd, P., and Dentler, W. 1999. Cytoplasmic dynein heavy chain 1b is required for flagellar assembly in Chlamydomonas. *Mol Biol Cell* **10**:693–712.

Preble, A.M., Giddings, T.H. Jr., and Dutcher, S.K. 2001. Extragenic bypass suppressors of mutations in the essential gene BLD2 promote assembly of basal bodies with abnormal microtubules in Chlamydomonas reinhardtii. *Genetics* **157**:163–181.

Randall, J. and Starling, D. 1972. Genetic determinant of the flagellum phenotype in Chlamydomonas reinhardtii. In *The Genetics of the Spermatozoon* (ed. R.A. Beatty and S. Gluecksohn-Waelsch Edinburgh). Scotland: University of Edinburgh, pp. 13–36.

Rieder, C.L. and Borisy, G.G. 1982. The centrosome cycle in PtK2 cells; asymmetric distribution and structural changes in the pericentriolar material. *Biol Cell* **44**:117–132.

Ringo, D.L. 1967. Flagellar motion and fine structure of the flagellar apparatus in Chlamydomonas. *J Cell Biol* **33**:543–571.

Rüffer, U. and Nultsch, W. 1987. Comparison of the beating of cis- and trans flagella of Chlamydomonas cells held on micropipettes. *Cell Motil* **7**:87–93.

Rüffer, U. and Nultsch, W. 1991. Flagellar photoresponses of Chlamydomonas cells held on micropipettes: II. Change in flagellar beat pattern. *Cell Motil Cytoskelton* **18**:269–278.

Rüffer, U. and Nultsch, W. 1998. Flagellar coordination in Chlamydomonas cells held on micropipettes. *Cell Motil Cytoskeleton* **41**:297–307.

Salisbury, J.L., Sanders, M.A., and Harpst, L. 1987. Flagellar root contraction and nuclear movement during flagellar regeneration in Chlamydomonas reinhardtii. *J Cell Biol* **105**:1799–1805.

Salisbury, J.L., Baron, A.T., and Sanders, M.A. 1988. The centrin-based cytoskeleton of Chlamydomonas reinhardtii: distribution in interphase and mitotic cells. *J Cell Biol* **107**:635–641.

Schermer, B., Hopker, K., Omran, H., Ghenoiu, C., Fliegauf, M., Fekete, A., Horvath, J., Kottgen, M., Hackl, M., Zschiedrich, S., Huber, T.B., Kramer-Zucker, A., Zentgraf, H., Blaukat, A., Walz, G., and Benzing, T. 2005. Phosphorylation by casein kinase 2 induces PACS-1 binding of nephrocystin and targeting to cilia. *EMBO J* **24**:4415–4424.

Schoppmeier, J. and Lechtreck, K.F. 2002. Localization of p210-related proteins in green flagellates and analysis of flagellar assembly in the green alga Dunaliella bioculatawith monoclonal anti-p210. *Protoplasma* **220**:29–38.

Silflow, C.D., LaVoie, M., Tam, L.W., Tousey, S., Sanders, M., Wu, W., Borodovsky, M., and Lefebvre, P.A. 2001. The Vfl1 protein in Chlamydomonas localizes in a rotationally asymmetric pattern at the distal ends of the basal bodies. *J Cell Biol* **153**:63–74.

Smyth, R.D. and Ebersold, W.T. 1985. Genetic investigation of a negatively phototactic strain of Chlamydomonas reinhardtii. *Genet Res* **46**:133–146.

Triemer, R.E. and Brown, R.M. Jr. 1976. Ultrastructure of meiosis in Chlamydomonas reinhardtii. *Br Phycol* **12**:23–44.

Vorobjev, I.A. and Chentsov, Y.S. 1982. Centrioles in the cell cycle. I. Epithelial cells. *J Cell Biol* **93**:938–949.

Weiss, R.L. 1984. Ultrastructure of the flagellar roots in Chlamydomonas gametes. *J Cell Sci* **67**:133–143.

Weiss, R.L., Goodenough, D.A., and Goodenough, U.W. 1977. Membrane differentiations at sites specialized for cell fusion. *J Cell Biol* **72**:144–160.

Wirschell, M., Hendrickson, T., and Sale, W.S. 2007. Keeping an eye on I1: I1 dynein as a model for flagellar dynein assembly and regulation. *Cell Motil Cytoskeleton* **64**:569–579.

Wloga, D., Rogowski, K., Sharma, N., Van Dijk, J., Janke, C., Edde, B., Bre, M.H., Levilliers, N., Redeker, V., Duan, J., Gorovsky, M.A., Jerka-Dziadosz, M., and Gaertig, J. 2008. Glutamylation on alpha-tubulin is not essential but affects the assembly and functions of a subset of microtubules in Tetrahymena thermophila. *Eukaryot Cell* **7**:1362–1372.

Yamashita, Y.M. and Fuller, M.T. 2008. Asymmetric centrosome behavior and the mechanisms of stem cell division. *J Cell Biol* **180**:261–266.

Yamashita, Y.M., Mahowald, A.P., Perlin, J.R., and Fuller, M.T. 2007. Asymmetric inheritance of mother versus daughter centrosome in stem cell division. *Science* **315**:518–521.

PROTEIN TRANSPORT IN AND OUT OF THE ENDOPLASMIC RETICULUM

TOM A. RAPOPORT

Howard Hughes Medical Institute and Department of Cell Biology, Harvard Medical School, Boston, Massachusetts

I. INTRODUCTION

A decisive step in the biosynthesis of many proteins is their partial or complete translocation across the eukaryotic endoplasmic reticulum (ER) membrane or the prokaryotic plasma membrane. Translocation occurs through a protein-conducting channel that is formed from a conserved, heterotrimeric membrane protein complex, the Sec61 or SecY complex. Structural and biochemical data suggest mechanisms that enable the channel to function with different partners, to open across the membrane, and to release laterally hydrophobic segments of membrane proteins into lipid. An ER translocation pathway in the reverse direction, called ERAD or retro-translocation, serves to destroy misfolded ER proteins in the cytosol. ERAD is still poorly understood, but many of the components involved have now been identified, setting the stage for mechanistic studies.

II. PROTEIN TRANSLOCATION INTO THE ER

Protein transport across the eukaryotic ER membrane is a decisive step in the biosynthesis of many proteins (for review, see Rapoport, 2007). These include soluble proteins, such as those ultimately secreted from the cell or localized to the ER lumen, and membrane proteins, such as those in the plasma membrane or in other organelles of the secretory pathway. Soluble proteins cross the membrane completely and usually have N-terminal, cleavable signal sequences, whose major feature is a segment of ~7–12 hydrophobic amino acids. Membrane proteins have different topologies, with one or more trans-membrane (TM) segments,

The Harvey Lectures, Series 102, pages 51–72
©2010 by John Wiley & Sons, Inc.

each containing ~20 hydrophobic amino acids; some hydrophilic regions of these proteins cross the membrane and others stay in the cytosol. Both types of proteins use the same machinery for translocation across the membrane: a protein-conducting channel. The channel allows polypeptides to cross the membrane and hydrophobic TM segments of membrane proteins to exit laterally into the lipid phase.

The ER membrane of eukaryotes is evolutionarily related to the plasma membrane of prokaryotes. Accordingly, in bacteria, secretory proteins cross directly the plasma membrane, employing signal sequences that are similar to those in eukaryotes. Membrane protein integration into the bacterial plasma membrane also occurs in a similar way as in the eukaryotic ER membrane. Because of these and other similarities, we will discuss protein translocation of these systems together.

The translocation channel is formed from an evolutionarily conserved heterotrimeric membrane protein complex, called the Sec61 complex in eukaryotes and the SecY complex in bacteria and archaea (Van den Berg et al., 2004). The α- and γ-subunits show significant sequence conservation, and both subunits are essential for the function of the channel and for cell viability. The β-subunits are not essential; they are similar in eukaryotes and archaea but show no obvious homology to the corresponding subunit in bacteria. The α-subunit forms the channel pore, as originally demonstrated by systematic cross-linking experiments (Mothes et al., 1994). Photoreactive probes incorporated into a polypeptide chain that was stalled during its translocation through the ER membrane gave cross-links only to the α-subunit of the Sec61 complex, suggesting that this protein surrounds the translocating chain during its membrane passage. In addition, reconstitution experiments showed that the Sec61/SecY complex is the only essential membrane component for protein translocation in mammals and bacteria (Brundage et al., 1990; Akimaru et al., 1991; Gorlich and Rapoport, 1993). The channel has an aqueous interior, as demonstrated by electrophysiology experiments and measurements of the fluorescence lifetime of probes incorporated into a translocating polypeptide chain (Simon and Blobel, 1991; Crowley et al., 1993, 1994).

III. Different Modes of Translocation

The channel alone is a passive pore; it must therefore associate with partners that provide a driving force for translocation. Depending on the channel partner, there are three known translocation modes (for review, see Rapoport, 2007). In co-translational translocation, the major partner is the ribosome. This translocation mode is found in all cells of all species. It is used for the translocation of secretory proteins, as well as for the integration of most membrane proteins. Co-translational translocation begins with a targeting phase, during which a ribosome-nascent chain complex is directed to the membrane. The signal or TM sequence of a growing polypeptide chain is recognized by the signal recognition particle (SRP), and the ribosome-SRP-nascent chain complex is then bound to the membrane, first by an interaction between SRP and its membrane receptor (SR), and then by an interaction between the ribosome and the translocation channel (Fig. 3.1a). The elongating poly-peptide chain subsequently moves directly from the tunnel inside the ribosome into the associated membrane channel; GTP hydrolysis during translation provides the energy for translocation. When the ribo-some synthesizes a cytosolic domain of a membrane protein, the polypep-tide chain emerges from the ribosome-channel junction sideways into the cytosol.

In most, if not all, cells some proteins are transported after their completion, i.e. post-translationally. This pathway is probably used only by soluble proteins, such as secretory proteins, that possess moderately hydrophobic signal sequences, allowing them to bypass recognition by the SRP. For these proteins to be transported, they need to remain unfolded or loosely folded after their release from the ribosome. Post-translational translocation occurs by different mechanisms in eukaryotes and bacteria. In yeast (and probably in all eukaryotes), translocation occurs by a ratcheting mechanism and involves as channel partners a membrane protein complex, the tetrameric Sec62/63 complex, and the luminal protein BiP, a member of the Hsp70 family of ATPases (Fig. 3.1b) (Matlack et al., 1999). BiP starts out in the ATP form with an open peptide-binding pocket. Following a transient interaction with the J-domain of Sec63p and ensuing ATP hydrolysis, the peptide-binding pocket closes around the translocation substrate. BiP is too large to fit through the channel and therefore prevents the bound polypeptide from sliding back into the cytosol. However, BiP does not prevent polypeptide movement in the forward direction. When moved sufficiently, the next BiP molecule binds, and this process is repeated until the entire polypep-tide chain is translocated. Finally, nucleotide exchange of ADP for ATP

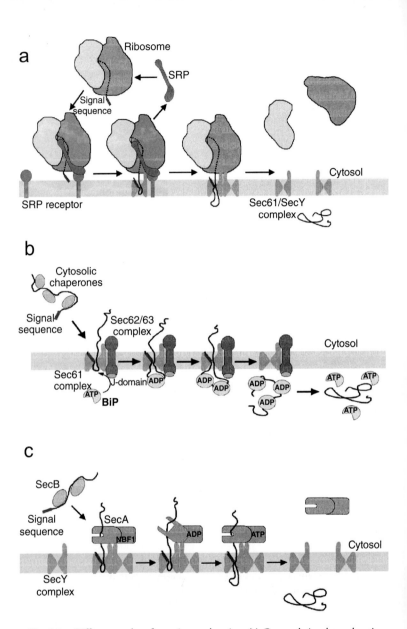

Fig. 3.1. Different modes of protein translocation. (a) Co-translational translocation. (b) Post-translational translocation in eukaryotes. It is conceivable that translocation is mediated by oligomers of the Sec61p complex, as in the other modes. (c) Post-translational translocation in bacteria.

opens the peptide-binding pocket and releases BiP from the translocated polypeptide.

In eubacterial post-translational translocation, polypeptides are "pushed" through the channel by its partner, the cytosolic ATPase SecA. SecA has two nucleotide-binding folds (NBF1 and 2), which bind the nucleotide between them and move relative to one another during the ATP hydrolysis cycle. Other domains are involved in polypeptide interaction. The translocation of many substrates begins with their binding to SecB, a cytosolic chaperone (Fig. 3.1c). Next, SecA interacts with SecB and accepts the polypeptide, probably binding both the signal sequence and the segment following it. The polypeptide chain is then transferred into the channel and translocated by a "pushing" mechanism (Economou and Wickner, 1994). A plausible mechanism assumes that a polypeptide-binding groove of SecA closes around the polypeptide chain and moves toward the channel, pushing the polypeptide into it. The size of SecA makes it unlikely that it inserts deeply into the SecY channel, as proposed earlier (Economou and Wickner, 1994). Upon nucleotide hydrolysis, the groove opens, releases the peptide, and moves away to "grab" the next segment of the substrate. This cycle continues until the entire polypeptide is translocated. Bacterial translocation in vivo requires an electrochemical gradient across the membrane, but the mechanism by which the gradient is utilized is unclear.

Archaea probably have both co- and post-translational translocation. While co-translational translocation is probably similar to that in eukaryotes and eubacteria, it is unknown how post-translational translocation occurs because archaea lack SecA, the Sec62/63 complex, and BiP.

IV. Structure and Function of the Translocation Channel

Much insight into the function of the channel came from the determination of the crystal structure, determined for an archaeal SecY complex at 3.2 Å resolution (Van den Berg et al., 2004). The structure is likely representative of all species, as indicated by sequence conservation and by the similarity to a lower-resolution structure of the *Escherichia coli* SecY complex determined by electron microscopy (EM) from two-dimensional (2-D) crystals (Breyton et al., 2002). Viewed from the cytosol, the channel has a square shape (Fig. 3.2a). The α-subunit is divided into two halves, TMs 1-5 and TMs 6-10. The loop between TM 5 and TM 6 at the back

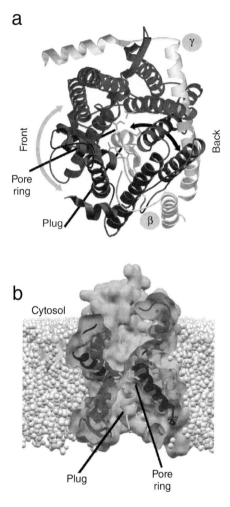

Fig. 3.2.   Structure of the translocation channel. (a) View from the cytosol of the X-ray structure of the SecY complex from *Methanococcus jannaschii*. The grey arrow indicates how the lateral gate opens. The black arrow indicates how the plug moves to open the channel across the membrane. (b) Cross section of the channel from the side.

of the α-subunit serves as a hinge, allowing the α-subunit to open at the front, the "lateral gate." The γ-subunit links the two halves of the α-subunit at the back by extending one TM segment diagonally across their interface. The β-subunit makes contact only with the periphery of the α-subunit, perhaps explaining why it is dispensable for the function of the complex.

The 10 helices of the α-subunit form an hourglass-shaped pore that consists of cytoplasmic and external funnels whose tips meet about halfway across the membrane (Fig. 3.2b). While the cytoplasmic funnel is empty, the external funnel is filled by a short helix, the "plug." The crystal structure represents a closed channel, but biochemical data indicate how it can open and translocate proteins (see below). The constriction of the hourglass-shaped channel is formed by a ring of six hydrophobic residues that project their side chains radially inward. The residues forming this "pore ring" are amino acids with bulky, hydrophobic side chains.

V. OPENING THE CHANNEL ACROSS THE MEMBRANE

The translocation of a secretory protein begins with its insertion into the channel. In all modes, the polypeptide inserts as a loop (Fig. 3.1a–c), with the signal sequence intercalated into the walls of the channel and the segment distal to it located in the pore proper. Opening of the channel for loop insertion occurs in two steps: (1) binding of a channel partner – the ribosome, the Sec62/63p complex, or SecA; and (2) intercalation of the hydrophobic segment of a signal sequence into the lateral gate, between TM2b and TM7, as indicated by photo-cross-linking experiments (Plath et al., 1998). The first step likely weakens interactions that keep the plug in the center of the Sec61/SecY molecule, as indicated by an increased ion conductance when non-translating ribosomes are bound to the channel (Simon and Blobel, 1991). The second step further destabilizes plug interactions, causing the plug to move from the center of Sec61/SecY into a cavity at the back of the molecule. Disulfide bridge cross-linking shows that the plug indeed comes close to the TM of the γ-subunit during translocation (Harris and Silhavy, 1999; Tam et al., 2005). This model is supported by the observation that many mutations that allow the translocation of proteins with defective signal sequences (signal suppressor mutations) appear to destabilize the closed channel. Finally, the open state of the channel is fixed by the insertion of the polypeptide segment distal to the signal sequence into the pore proper. During subsequent translocation, the signal sequence remains stationary, while the rest of the polypeptide moves through the pore. The plug can only return to the center of Sec61/SecY when the polypeptide chain has left the pore.

VI. THE PORE

The crystal structure indicates that a single copy of the Sec61/SecY complex forms the pore (Van den Berg et al., 2004); a polypeptide would move from the cytoplasmic funnel through the pore ring into the external funnel. Previously, it was thought that several copies of the Sec61/SecY complex would assemble to form a hydrophilic pore in the membrane, but the crystal structure shows that, similar to all other membrane proteins, a single SecY complex has an entirely hydrophobic belt around its exterior surface. Systematic disulfide cross-linking experiments, in which one cysteine is placed in a translocation substrate and another at various positions of SecY, show that the polypeptide chain indeed passes through the center of the SecY complex and makes contact with it only at the waist of the hourglass-shaped channel (Cannon et al., 2005). The aqueous interior of the channel and its shape help to minimize the energy required to move a translocation substrate through the membrane.

The diameter of the pore ring as observed in the crystal structure is too small to allow the passage of most polypeptide chains. Pore widening, therefore, has to be postulated, which could be mediated by movements of the helices to which the pore ring residues are attached. The flexibility of the pore ring is supported by molecular dynamics simulations as well as by electrophysiology experiments (Gumbart and Schulten, 2006; Saparov et al., 2007). The intercalation of a signal sequence into the walls of the channel may cause additional widening of the pore, as is required for loop insertion of a polypeptide chain. The estimated maximum dimensions of the pore, based on the crystal structure, would allow a polypeptide to form an α-helix, but no tertiary structure, in agreement with experimental data.

Fluorescence quenching experiments suggested that the pore is much larger (40–60 Å) than conceivable with a single Sec61/SecY molecule (Hamman et al., 1997). These data could be reconciled with the crystal structure if two or more Sec61/SecY complexes associated at their front surfaces and opened their lateral gates to fuse their pores into a larger channel. Although a different arrangement – a back-to-back orientation – is suggested by the 2-D structure of the *E. coli* SecY complex and by cross-linking data (Kaufmann et al., 1999; Breyton et al., 2002), the model would be consistent with an EM structure in which a translating *E. coli* ribosome was proposed to be associated with two nearly front-to-front orientated SecY molecules (Mitra et al., 2005). However, this structure is based on a low-resolution electron density map (~15 Å), and the docking of the SecY crystal structure required its drastic modification. The position and orientation of both SecY molecules are different from that of the single SecY molecule that is seen in more recent EM structures of non-translating ribosome-SecY complexes (Menetret et al., 2007). Disulfide bridge cross-linking experiments argue against fusion of different pores because they show that, during SecA-mediated translocation, both the signal sequence and the mature region of a polypeptide chain are located in the same SecY molecule (Osborne and Rapoport, 2007). A detergent-solubilized translocation intermediate also contains just one copy of SecY associated with one SecA and one translocation substrate molecule (Duong, 2003). It should be noted that in the fluorescence quenching experiments, the fluorescent probes were located deep inside the ribosome, and therefore the same large diameter (40–60 Å) must be assumed for the ribosome tunnel, a size that does not agree with that seen in ribosome structures (<20 Å) determined by crystallography or cryo-EM. It is also difficult to see how the rigid ribosomal RNA that lines most of the tunnel could undergo such a dramatic change. It has been argued that the structural methods give inaccurate answers because they are obtained in detergent. However, the disulfide cross-linking experiments that showed that both the signal sequence and the mature region of a polypeptide are in the same SecY molecule were performed with functional translocation intermediates in intact membranes (Osborne and Rapoport, 2007).

VII. Oligomeric Translocation Channels

Although the pore is formed by only one Sec61/SecY molecule, translocation of a polypeptide chain appears to be mediated by oligomers. This conclusion is based on the observation that a SecY molecule defective in SecA-mediated translocation can be rescued by linking it covalently with a wild-type SecY copy (Osborne and Rapoport, 2007). Disulfide bridge cross-linking showed that SecA binds through its NBF1 domain to a non-translocating SecY copy and moves the polypeptide chain through a neighboring SecY copy. The static interaction with the non-translocating SecY molecule would prevent complete detachment of SecA when its peptide-binding domain moves away from the translocating SecY molecule to "grab" the next polypeptide segment.

The Sec61/SecY complex probably forms oligomers during co-translational translocation as well. When a ribosome/nascent chain/SRP complex binds to the SRP receptor, a domain of SRP undergoes a conformational change that exposes a site on the ribosome to which a single Sec61/SecY molecule could bind (Halic et al., 2006), likely the one seen in recent EM structures of non-translating ribosome-SecY complexes (Menetret et al., 2007). The bound SecY molecule is close to the point where a polypeptide exits the ribosome and could thus become the translocating copy. At a later stage of translocation, SRP completely detaches from the ribosome, and one or more additional copies of the Sec61/SecY complex may associate (Fig. 3.1a), as suggested by cross-linking and freeze-fracture EM experiments (Hanein et al., 1996; Schaletzky and Rapoport, 2006). These copies could stabilize the ribosome-channel junction and possibly recruit other components, such as signal peptidase and oligosaccharyl transferase, or the translocon-associated protein complex (TRAP). Upon termination of translocation, dissociation of the Sec61/SecY oligomers could facilitate the release of the ribosome from the membrane. Dissociable oligomers may also allow the Sec61/SecY complex to change channel partners and modes of translocation.

EM structures of detergent-solubilized ribosome-channel complexes suggested the presence of three or four Sec61 molecules (Beckmann et al., 2001; Menetret et al., 2005). However, the low resolution of these structures makes it difficult to distinguish between protein and additional density contributed by lipid and detergent, and it is therefore possible that only one Sec61 molecule is present, while the other Sec61 copies were lost during solubilization, similar to the dissociation seen with SecA-interacting SecY oligomers.

The emerging concept of homo-oligomeric channels, in which only one copy is active at any given time, could apply to other protein translocation systems, such as for PapC, involved in the secretion of pili subunits across the outer membrane of *E. coli* (Thanassi et al., 2005), or for Tom40 and Tim22, involved in protein transport across the outer and inner mitochondrial membrane, respectively (Ahting et al., 1999; Rehling et al., 2003).

VIII. Membrane Protein Integration

During the synthesis of a membrane protein, hydrophobic TMs exit the channel laterally, moving from the aqueous interior of the channel through the lateral gate into the lipid phase. The gate is formed by short segments of four TMs at the front of Sec61/SecY (Van den Berg et al., 2004). The resulting seam in the wall of the channel is probably weak once the plug has moved toward the back of the channel and no longer contacts these TM segments. The lateral gate may then continuously open and close, exposing polypeptide segments located in the aqueous channel to the surrounding hydrophobic lipid phase. Segments that are sufficiently long and hydrophobic would exit the channel through the lateral gate simply by partitioning between aqueous and hydrophobic environments. This model is supported by photo-cross-linking experiments (Heinrich et al., 2000) and by the agreement between a hydrophobicity scale derived from peptide partitioning into an organic solvent and the tendency of a peptide to span the membrane (Hessa et al., 2005). The size of the channel indicates that TMs exit laterally one by one or in pairs. Hydrophilic segments between the TMs would alternately move from the ribosome through the aqueous channel to the external side of the membrane, or emerge between the ribosome and channel into the cytosol. Movement into the cytosol likely uses a "gap" between the ribosome and channel, which can be visualized in EM structures (Beckmann et al., 2001; Menetret et al., 2005).

In contrast to signal sequences, which always have their N-termini in the cytosol, the first TM segment of a membrane protein can have its N-terminus on either side of the membrane, depending on the amino acid sequence of the protein. In a multi-spanning membrane protein, the first TM often determines the orientation of the subsequent ones, although there are exceptions in which internal TMs have a preferred orientation regardless of the behavior of preceding TMs. If the first hydrophobic segment is long and the N-terminus is not retained in the cytosol by positive charges or by the folding of the preceding polypeptide segment, it can flip across the channel and, subsequently, exit it laterally into the lipid phase. If the N-terminus is retained in the cytosol and the polypeptide chain is further elongated, the C-terminus can translocate across the channel, inserting the polypeptide as a loop, as in the case of a secretory protein.

IX. Maintaining the Permeability Barrier

The channel must prevent the free movement of small molecules, such as ions or metabolites. Maintaining the membrane barrier is particularly important for prokaryotes, as the proton gradient across the membrane is the main source of their energy, but even the eukaryotic ER membrane must prevent the free flow of ions.

A complex molecular mechanism to maintain the membrane permeability barrier is suggested by fluorescence quenching experiments with ER membranes. In this model, the resting channel has a pore size of 9–15 Å, which is closed at the luminal end by BiP (Hamman et al., 1998). During the translocation of a secretory protein, the channel widens to 40–60 Å, the luminal seal is lost, and instead a cytoplasmic seal is formed by the translating ribosome. During the synthesis of a multi-spanning membrane protein, the seals provided by the ribosome and BiP alternate, depending on whether the nascent chain is directed to the ER lumen or the cytosol. The pore is closed by BiP before the TM in a nascent chain reaches the channel, implying that the ribosome recognizes the nascent chain as a membrane protein. Although TMs can form α-helices inside the ribosome, it is difficult to see how the ribosome tunnel, with its mostly hydrophilic surface, could recognize the hydrophobic sequence of a TM. In addition, consecutive TMs move in the same direction inside the ribosome but would have to transmit opposing signals to the ribosome-associated channel. A tight seal between the ribosome and channel is also at odds with EM structures that reveal a gap of 12–15 Å between them (Beckmann et al., 2001; Menetret et al., 2005). Finally, this model does not explain how the membrane barrier is maintained in the absence of a ribosome (in post-translational translocation) or in the absence of BiP (in prokaryotes).

The crystal structure suggests a simpler model, in which the membrane barrier is formed by the channel itself, with both the plug and pore ring contributing to the seal. Electrophysiology experiments show that the resting SecY channel, in the absence of other components, is indeed impermeable to ions and water, and opens upon plug displacement (Saparov et al., 2007). In the active channel, the passage of small molecules might be restricted by the pore ring fitting like a gasket around the translocating polypeptide chain. The seal would not be expected to be perfect, but leakage could be compensated for by powerful ion pumps. During the synthesis of a multi-spanning membrane protein, the seal would be provided in an alternating manner by either the nascent chain in the pore, or – once the chain has left the pore – by the plug returning to the center of Sec61/SecY. This model needs further experimental verification, but it would explain how the membrane barrier can be maintained in both co- and post-translational translocation, and why a gap between the ribosome and channel may not compromise the barrier.

Surprisingly, plug deletion mutants are viable in *Saccharomyces cerevisiae* and *E. coli* and have only moderate translocation defects (Junne et al., 2006; Li et al., 2007; Maillard et al., 2007). However, the crystal structure of these mutants shows that new plugs are formed from neighboring polypeptide segments (Li et al., 2007). The new plugs still seal the closed channel, but they have lost many interactions that normally keep the plug in the center of SecY. This results in continuous channel opening and closing, and permits polypeptides with defective or even missing signal sequences to be translocated. The plug sequences are only poorly conserved among Sec61/SecY channels, supporting the idea that promiscuous segments can seal the channel and lock it in its closed state.

X. Transport of Proteins Out of the ER

Some proteins are translocated across or integrated into the ER membrane but cannot reach their native conformation; they are often transported back into the cytosol and destroyed by the proteasome (for review, see Tsai et al., 2002; Meusser et al., 2005). This process is called ERAD (for ER-associated degradation), retro-translocation, or dislocation. The pathway is hijacked by certain viruses, such as the human cytomegalovirus (Lilley and Ploegh, 2005). This virus codes for two proteins, US2 and US11, which both trigger the retro-translocation of the Major Histocompatibility Complex (MHC) class I heavy chain into the cytosol. This process prevents the presentation of viral peptides on the surface of virus-infected cells and allows the virus to be propagated without the cells being destroyed by cytotoxic T cells. The ERAD pathway is also hijacked by some plant and bacterial toxins. The best studied example is cholera toxin (Lord et al., 2005). This toxin binds through its B-subunits to the surface of intestinal cells, and travels backwards along the secretory pathway to the ER. Then a fragment of the A chain (A1) crosses the ER membrane into the cytosol, utilizing the cellular ERAD machinery. Once in the cytosol, the A1 chain refolds, becomes an enzyme that adenoribosylates a trimeric G protein at the plasma membrane, which in turn opens chloride channels, resulting in massive chloride and water secretion.

The mechanism of ERAD is still poorly understood, but it comprises several different steps. The first step is the recognition of an ERAD substrate, be it a soluble, luminal protein or a membrane-bound protein. Next, the substrate is targeted to the retro-translocation machinery. At least in the case of luminal proteins, it is likely that a retro-translocation channel is required to move proteins through the membrane. In the case of membrane proteins, it is conceivable that they are pulled out of the membrane without the involvement of a channel. Once in the cytosol, most proteins are poly-ubiquitinated and then delivered to the proteasome for degradation.

Early steps in the retro-translocation of cholera toxin were elucidated with an in vitro system, in which cholera toxin was incubated with ER luminal extracts (Tsai et al., 2001). These studies showed that protein disulfide isomerase (PDI) functions to disassemble and unfold the toxin, once its A-chain has been cleaved into A1 and A2 chains. PDI acts as a redox-driven chaperone: in the reduced state, it binds to the A1-chain and in the oxidized state it releases it. The complex of reduced PDI and toxin binds to an unknown membrane protein at the luminal side of the ER membrane. Subsequently, the peptide-binding pocket of PDI is opened by oxidation, catalyzed by the enzyme Ero1p (Tsai and Rapoport, 2002). PDI and related proteins may play a general role in initiating retro-translocation of proteins to the cytosol.

Late stages of retro-translocation were elucidated with the US11-dependent degradation of MHC class I heavy chains. A permeabilized system was developed in which US11-expressing cells were treated with a low concentration of digitonin, thus allowing cytosolic proteins to be released (Shamu et al., 1999). This approach led to the identification of cytosolic proteins involved in retro-translocation. Using AMPPNP as an ATP analog, it was shown that the MHC class I heavy chains were still poly-ubiquitinated (the ubiquitin-activating enzyme cleaves ATP into

AMP and PPi) but not moved from the membrane into the cytosol (Flierman et al., 2003). The ATPase required for this step turned out to be a member of the AAA family of ATPases, called p97 or VCP in mammalian cells and Cdc48p in yeast (Ye et al., 2001). The ATPase consists of an N-terminal N domain and two ATPase domains, called D1 and D2. Six D1-D2 domains form a double-barrel structure. The N-domains sit on the side of the D1 ring and bind a conserved cofactor, consisting of the proteins Ufd1p and Npl4p. The ATPase complex appears to initially recognize unmodified segments of a substrate, presumably because they are unfolded (Ye et al., 2003). Once a poly-ubiquitin chain has been attached to a substrate, it is recognized by both the N-domain and Ufd1p in a synergistic manner. The binding of poly-ubiquitin might activate the p97/Cdc48 ATPase to "pull" polypeptides out of the ER membrane. The ATPase complex is required for all ERAD substrates that are poly-ubiquitinated. Only two ERAD substrates are known not to be poly-ubiquitinated (prepro-α-factor in yeast and cholera toxin in mammalian cells), and these do not require the ATPase complex (McCracken and Brodsky, 2003; Kothe et al., 2005).

A membrane receptor complex was identified in mammalian cells by adding recombinant p97 protein to ER membranes, solubilization in detergent, and pulling on the ATPase (Ye et al., 2004). Two integral membrane proteins were found to copurify with the ATPase. One is Derlin-1, a homolog of yeast Der1, which had been identified previously in yeast in a screen for components involved in the degradation of a misfolded ER protein (Knop et al., 1996). Derlin-1 was also found as an interaction partner of US11 (Lilley and Ploegh, 2004; Ye et al., 2004). The other protein is VIMP, a component that does not have obvious homologs in lower organisms. Later studies showed that the ATPase also binds to the ubiquitin ligases Hrd1p and gp78 and, at least in yeast, to an adaptor protein, called Ubx2 (Schuberth et al., 2004; Neuber et al., 2005). Together, these interactions bring into close proximity all components that are required for late stages of retro-translocation.

A systematic analysis of interactions in yeast resulted in the identification of most, if not all, ERAD components (Carvalho et al., 2006). Using affinity tags on known ERAD components, all interaction partners could be identified that remain associated upon detergent solubilization. These results and previous experiments (Vashist and Ng, 2004) led to the concept that there are three distinct ERAD pathways: ERAD-L for the degradation

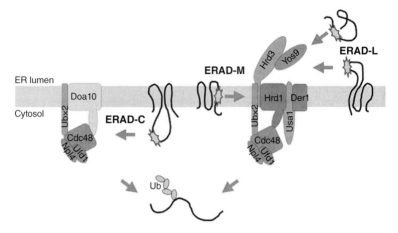

Fig. 3.3.   Three different ERAD pathways for the degradation of ER proteins in yeast. The location of the misfolded domain determines which pathway is used (ERAD-L, ERAD-M, or ERAD-C). The pathways use different ubiquitin ligases but converge at the point where the cytoplasmic Cdc48 ATPase complex functions.

of proteins with misfolded luminal domains (these can be soluble or membrane bound), ERAD-M for the degradation of proteins with misfolded intra-membrane domains, and ERAD-C for the degradation of membrane proteins with misfolded cytosolic domains. Each of these pathways requires a distinct ubiquitin-ligase complex, but they all converge at the point where the cytosolic ATPase complex (Cdc48, Ufd1, Npl4) comes into play (the membrane adaptor protein Ubx2 is also common to all pathways). Upstream of the ATPase complex, ERAD-C requires the ubiquitin ligase Doa10p, and ERAD-M and ERAD-L require the ligase Hrd1p. Hrd1p is in a complex with three other membrane proteins (Hrd3p, Usa1p, Der1p), but the latter two are dispensable for ERAD-M. ERAD-L requires on the luminal side of the ER the lectin-like protein Yos9p, which interacts with Hrd3p and may transfer substrates to the Hrd1p complex. All identified yeast components have at least one homolog in mammals, suggesting that the basic principles of ERAD may be the same in all eukaryotes. However, some components have more than one homolog in mammals, perhaps each dedicated to a subset of substrates.

An unresolved issue is the nature of the retro-translocation channel, which likely is required for at least soluble ERAD-L substrates. One possibility is that retro-translocation occurs through the same Sec61p channel that is used for the transport of proteins in the "forward" direction, from the cytosol into the ER. However, the experiments implicating Sec61p in ERAD are not entirely convincing (for a discussion, see Tsai et al., 2002), and it is difficult to see how luminal substrates could move the plug and open the channel. Other candidates for the retro-translocation channel have been proposed, including the ubiquitin ligase Hrd1p (Gauss et al., 2006) and Der1p (Lilley and Ploegh, 2004; Ye et al., 2004), which are both multi-spanning membrane proteins.

XI. PERSPECTIVE

We are beginning to understand protein translocation across the eukaryotic ER and bacterial plasma membranes at the molecular level. In particular, progress during the past several years has led to important insight into the function of the Sec61/SecY channel. Nevertheless, major questions in the field are still controversial and unresolved, and further progress will require the combination of different approaches. Electro-physiology experiments are needed to complement the fluorescence quenching method, particularly because results from the latter are difficult to reconcile with structural data. Important questions in co-translational translocation include how the SRP receptor and channel collaborate, how many Sec61/SecY complexes participate in translocation, and how the ribosome ultimately dissociates from the channel. The precise role of the Sec62/63 components in post-translational translocation, and the mechanism by which SecA moves polypeptides also need to be clarified. Membrane protein integration is still particularly poorly understood, and new methods are required to follow the membrane integration of TMs. Several other translocation components have been identified, such as the TRAM protein and the TRAP complex in mammalian cells, or the YidC and SecDF proteins in prokaryotes. These components may be required as chaperones for the folding of TM segments, or to increase the efficiency of translocation of some substrates, but their precise functions remain to be clarified. Much of the progress in the field will hinge on structural data, with the "holy grail" being a picture of the "translocon in action," where a channel associated with both a partner and a translocating poly-peptide chain is visualized at atomic detail.

ERAD still lacks much behind "forward translocation." Arguably, the most important unresolved issue is the nature of the retro-translocation channel. A major goal must be the establishment of defined translocation intermediates, analogous to those that were so instrumental in identifying the Sec61p channel. None of the genetic screens was exhaustive, and protein complexes may have dissociated during the detergent solubilization of membranes, so it is conceivable that important components, such as the channel, may not yet have been identified. But even if the channel component is among the identified proteins (the most likely candidates being Hrd1p/gp78 and Der1p/Derlin), direct evidence needs to be provided. In addition, one would like to know the role of the other components in ERAD. The ultimate goal must be a reconstituted system with purified components, which reproduces ERAD in vitro.

Both "forward" and retro-translocation remain exciting areas of research, with very different levels of approaches and understanding. The results will likely serve as a paradigm for other protein translocation systems, such as those in mitochondria, chloroplasts, and peroxisomes. In the case of ERAD, they will have important implications for diseases in which proteins misfold in the ER, and therefore cannot perform their normal functions in the cell.

Acknowledgment

Work in the author's laboratory has been supported by grants from the National Institute of Health. The author is a Howard Hughes Medical Institute Investigator.

References

Ahting, U., Thun, C., Hegerl, R., Typke D., Nargang, F.E., Neupert, W., and Nussberger, S. 1999. The TOM core complex: the general protein import pore of the outer membrane of mitochondria. *J Cell Biol* **147**:959–968.

Akimaru, J., Matsuyama, S., Tokuda, H., and Mizushima, S. 1991. Reconstitution of a protein translocation system containing purified SecY, SecE, and SecA from Escherichia coli. *Proc Natl Acad Sci U S A* **88**:6545–6549.

Beckmann, R., Spahn, C.M., Eswar, N., Helmers, J., Penczek, P.A., Sali, A., Frank, J., and Blobel, G. 2001. Architecture of the protein-conducting channel associated with the translating 80S ribosome. *Cell* **107**:361–372.

Breyton, C., Haase, W., Rapoport, T.A., Kuhlbrandt, W., and Collinson, I. 2002. Three-dimensional structure of the bacterial protein-translocation complex SecYEG. *Nature* **418**:662–665.

Brundage, L., Hendrick, J.P., Schiebel, E., Driessen, A.J., and Wickner, W. 1990. The purified E. coli integral membrane protein SecY/E is sufficient for reconstitution of SecA-dependent precursor protein translocation. *Cell* **62**: 649–657.

Cannon, K.S., Or, E., Clemons, W.M. Jr., Shibata, Y., and Rapoport, T.A. 2005. Disulfide bridge formation between SecY and a translocating polypeptide localizes the translocation pore to the center of SecY. *J Cell Biol* **169**: 219–225.

Carvalho, P., Goder, V., and Rapoport, T.A. 2006. Distinct ubiquitin-ligase complexes define convergent pathways for the degradation of ER proteins. *Cell* **126**:361–373.

Crowley, K.S., Reinhart, G.D., and Johnson, A.E. 1993. The signal sequence moves through a ribosomal tunnel into a noncytoplasmic aqueous environment at the ER membrane early in translocation. *Cell* **73**:1101–1115.

Crowley, K.S., Liao, S.R., Worrell, V.E., Reinhart, G.D., and Johnson, A.E. 1994. Secretory proteins move through the endoplasmic reticulum membrane via an aqueous, gated pore. *Cell* **78**:461–471.

Duong, F. 2003. Binding, activation and dissociation of the dimeric SecA ATPase at the dimeric SecYEG translocase. *EMBO J* **22**:4375–4384.

Economou, A. and Wickner, W. 1994. SecA promotes preprotein translocation by undergoing ATP-driven cycles of membrane insertion and deinsertion. *Cell* **78**:835–843.

Flierman, D., Ye, Y., Dai, M., Chau, V., and Rapoport, T.A. 2003. Polyubiquitin serves as a recognition signal, rather than a ratcheting molecule, during retrotranslocation of proteins across the endoplasmic reticulum membrane. *J Biol Chem* **278**:34774–34782.

Gauss, R., Sommer, T., and Jarosch, E. 2006. The Hrd1p ligase complex forms a linchpin between ER-lumenal substrate selection and Cdc48p recruitment. *EMBO J* **25**:1827–1835.

Gorlich, D. and Rapoport, T.A. 1993. Protein translocation into proteoliposomes reconstituted from purified components of the endoplasmic reticulum membrane. *Cell* **75**:615–630.

Gumbart, J. and Schulten, K. 2006. Molecular dynamics studies of the archaeal translocon. *Biophys J* **90**:2356–2367.

Halic, M., et al. 2006. Signal recognition particle receptor exposes the ribosomal translocon binding site. *Science* **312**:745–747.

Hamman, B.D., Chen, J.C., Johnson, E.E., and Johnson, A.E. 1997. The aqueous pore through the translocon has a diameter of 40-60Å during cotranslational protein translocation at the ER membrane. *Cell* **89**:535–544.

Hamman, B.D., Hendershot, L.M., and Johnson, A.E. 1998. BiP maintains the permeability barrier of the ER membrane by sealing the lumenal end of the translocon pore before and early in translocation. *Cell* **92**: 747–758.

Hanein, D., Matlack, K.E., Jungnickel, B., Plath, K., Kalies, K.V., Miller, K.R., Rapoport, T.A., and Akey, C.W. 1996. Oligomeric rings of the Sec61p complex induced by ligands required for protein translocation. *Cell* **87**: 721–732.

Harris, C.R. and Silhavy, T.J. 1999. Mapping an interface of SecY (PrlA) and SecE (PrlG) by using synthetic phenotypes and in vivo cross-linking. *J Bacteriol* **181**:3438–3444.

Heinrich, S.U., Mothes, W., Brunner, J., and Rapoport, T.A. 2000. The Sec61p complex mediates the integration of a membrane protein by allowing lipid partitioning of the transmembrane domain. *Cell* **102**:233–244.

Hessa, T., Kim, H., Bihlmaier, K., Lundin, C., Boekel, J., Andersson, H., Nilsson, I., White, S.H., and von Heijne, G. 2005. Recognition of transmembrane helices by the endoplasmic reticulum translocon. *Nature* **433**:377–381.

Junne, T., Schwede, T., Goder, V., and Spiess, M. 2006. The plug domain of yeast Sec61p is important for efficient protein translocation, but is not essential for cell viability. *Mol Biol Cell* **17**:4063–4068.

Kaufmann, A., Manting, E.H., Veenendaal, A.K., Driessen, A.J., and van der Does, C. 1999. Cysteine-directed cross-linking demonstrates that helix 3 of SecE is close to helix 2 of SecY and helix 3 of a neighboring SecE. *Biochemistry* **38**:9115–9125.

Knop, M., Finger, A., Braun, T., Hellmuth, K., and Wolf, D.H. 1996. Der1, a novel protein specifically required for endoplasmic reticulum degradation in yeast. *EMBO J* **15**:753–763.

Kothe, M., Ye, Y., Wagner, J.S., DeLuca, H.E., Kern, E., Rapoport, T.A., and Lencer, W.I. 2005. Role of p97 AAA-ATPase in the retrotranslocation of the cholera toxin A1 chain, a non-ubiquitinated substrate. *J Biol Chem* **280**:28127–28132.

Li, W., Schulman, S., Boyd, D., Erlandson, K., Beckwith, J., and Rapoport, T.A. 2007. The plug domain of the SecY protein stabilizes the closed state of the translocation channel and maintains a membrane seal. *Mol Cell* **26**: 511–521.

Lilley, B.N. and Ploegh, H.L. 2004. A membrane protein required for dislocation of misfolded proteins from the ER. *Nature* **429**:834–840.

Lilley, B.N. and Ploegh, H.L. 2005. Viral modulation of antigen presentation: manipulation of cellular targets in the ER and beyond. *Immunol Rev* **207**: 126–144.

Lord, J.M., Roberts, L.M., and Lencer, W.I. 2005. Entry of protein toxins into mammalian cells by crossing the endoplasmic reticulum membrane: co-opting basic mechanisms of endoplasmic reticulum-associated degradation. *Curr Top Microbiol Immunol* **300**:149–168.

Maillard, A.P., Lalani, S., Silva, F., Belin, D., and Duong, F. 2007. Deregulation of the SecYEG translocation channel upon removal of the plug domain. *J Biol Chem* **282**:1281–1287.

Matlack, K.E., Misselwitz, B., Plath, K., and Rapoport, T.A. 1999. BiP acts as a molecular ratchet during posttranslational transport of prepro-alpha factor across the ER membrane. *Cell* **97**:553–564.

McCracken, A.A. and Brodsky, J.L. 2003. Evolving questions and paradigm shifts in endoplasmic-reticulum-associated degradation (ERAD). *Bioessays* **25**:868–877.

Menetret, J.F., Hegde, R.S., Heinrich, S.U., Chandramouli, P., Ludtke, S.J., Rapoport, T.A., and Akey, C.W. 2005. Architecture of the ribosome-channel complex derived from native membranes. *J Mol Biol* **348**:445–457.

Menetret, J.F., Schaletzky, J., Clemons, W.M., Jr., Osborne, A.R., Skånland, S.S., Denison, C., Gygli, S.P., Kirkpatrick, D.S., Park, E., Ludtke, S.J., Rapoport, T.A., and Akey, C.W. 2007. Ribosome binding of a single copy of the SecY complex: implications for the initiation of protein translocation. *Mol Cell*.

Meusser, B., Hirsch, C., Jarosch, E., and Sommer, T. 2005. ERAD: the long road to destruction. *Nat Cell Biol* **7**:766–772.

Mitra, K., Schaffitzd, C., Shaikh, T., Tama, F., Jenni, S., Brooks, C.L., 3rd, Ban, N., and Frank, J. 2005. Structure of the E. coli protein-conducting channel bound to a translating ribosome. *Nature* **438**:318–324.

Mothes, W., Prehn, S., and Rapoport, T.A. 1994. Systematic probing of the environment of a translocating secretory protein during translocation through the ER membrane. *EMBO J* **13**:3937–3982.

Neuber, O., Jarosch, E., Volkwein, C., Walter, J., and Sommer, T. 2005. Ubx2 links the Cdc48 complex to ER-associated protein degradation. *Nat Cell Biol* **7**:993–998.

Osborne, A.R. and Rapoport, T.A. 2007. Protein translocation is mediated by oligomers of the SecY complex with one SecY copy forming the channel. *Cell* **129**:97–110.

Plath, K., Mothes, W., Wilkinson, B.M., Stirling, C.J., and Rapoport, T.A. 1998. Signal sequence recognition in posttranslational protein transport across the yeast ER membrane. *Cell* **94**:795–807.

Rapoport, T.A. 2007. Protein translocation across the eukaryotic endoplasmic reticulum and bacterial plasma membranes. *Nature* **450**:663–669.

Rehling, P., Model, K., Brandner, K., Kovermann, P., Sickmann, A., Meyer, H.E., Kühlbrandt, W., Wagner, R., Truscott, K.N., and Pfanner, N. 2003. Protein insertion into the mitochondrial inner membrane by a twin-pore translocase. *Science* **299**:1747–1751.

Saparov, S.M., Erlandson, K., Cannon, K., Schaletzky, J., Scholman, S., Rapoport, T.A., and Pohl, P. 2007. Determining the conductance of the SecY protein translocation channel for small molecules. *Mol Cell* **26**:501–509.

Schaletzky, J. and Rapoport, T.A. 2006. Ribosome binding to and dissociation from translocation sites of the endoplasmic reticulum membrane. *Mol Biol Cell* **17**:3860–3869.

Schuberth, C., Richly, H., Rumpf, S., and Buchberger, A. 2004. Shp1 and Ubx2 are adaptors of Cdc48 involved in ubiquitin-dependent protein degradation. *EMBO Rep* **5**:818–824.

Shamu, C.E., Story, C.M., Rapoport, T.A., and Ploegh, H.L. 1999. The pathway of US11-dependent degradation of MHC class I heavy chains involves a ubiquitin-conjugated intermediate. *J Cell Biol* **147**:45–58.

Simon, S.M. and Blobel, G. 1991. A protein-conducting channel in the endoplasmic reticulum. *Cell* **65**:371–380.

Tam, P.C., Maillard, A.P., Chan, K.K., and Duong, F. 2005. Investigating the SecY plug movement at the SecYEG translocation channel. *EMBO J* **24**:3380–3388.

Thanassi, D.G., Stathopoulos, C., Karkal, A., and Li, H. 2005. Protein secretion in the absence of ATP: the autotransporter, two-partner secretion and chaperone/usher pathways of gram-negative bacteria (review). *Mol Membr Biol* **22**:63–72.

Tsai, B. and Rapoport, T.A. 2002. Unfolded cholera toxin is transferred to the ER membrane and released from protein disulfide isomerase upon oxidation by Ero1. *J Cell Biol* **159**:207–216.

Tsai, B., Rodighiero, C., Lencer, W.I., and Rapoport, T.A. 2001. Protein disulfide isomerase acts as a redox-dependent chaperone to unfold cholera toxin. *Cell* **104**:937–948.

Tsai, B., Ye, Y., and Rapoport, T.A. 2002. Retro-translocation of proteins from the endoplasmic reticulum into the cytosol. *Nat Rev Mol Cell Biol* **3**:246–255.

Van den Berg, B., Clemons, W.M., Jr., Collinson, I., Modis, Y., Hartmann, E., Harrison, S.C., and Rapoport, T.A. 2004. X-ray structure of a protein-conducting channel. *Nature* **427**:36–44.

Vashist, S. and Ng, D.T. 2004. Misfolded proteins are sorted by a sequential checkpoint mechanism of ER quality control. *J Cell Biol* **165**:41–52.

Ye, Y., Meyer, H.H., and Rapoport, T.A. 2001. The AAA ATPase Cdc48/p97 and its partners transport proteins from the ER into the cytosol. *Nature* **414**:652–656.

Ye, Y., Meyer, H.H., and Rapoport, T.A. 2003. Function of the p97-Ufd1-Npl4 complex in retrotranslocation from the ER to the cytosol: dual recognition of nonubiquitinated polypeptide segments and polyubiquitin chains. *J Cell Biol* **162**:71–84.

Ye, Y., Shibata, Y., Yun, C., Ron, D., and Rapoport, T.A. 2004. A membrane protein complex mediates retro-translocation from the ER lumen into the cytosol. *Nature* **429**:841–847.

SIGNALING NETWORKS THAT CONTROL SYNAPSE DEVELOPMENT AND COGNITIVE FUNCTION

MICHAEL E. GREENBERG

Children's Hospital Boston, Program in Neurobiology, Harvard Medical School, Department of Neurobiology, Boston, Massachusetts

I. INTRODUCTION

Over 40 years ago, Drs. David Hubel and Thorsten Wiesel demonstrated in the mammalian visual system that our interactions with the world affect the development of our brains (Wiesel and Hubel, 1963). This process is initiated by environmental stimuli (e.g. visual stimuli) and is manifested by the release of neurotransmitter at synapses in the brain that affect the postsynaptic neuron and leads to a remodeling of the synapses and neural circuits that mediate the organism's response to the stimulus. These experience-dependent changes in synaptic structure and function must take place in order to create a correctly wired brain and are essential to learning and memory throughout life. Thus, it is clear that in addition to the hardwired genetic program that directs the development of an organism, experience plays a key role in shaping the development of the nervous system. Given that experience through neuronal activity shapes the circuitry of the nervous system, it follows that defects in this process, whether genetic defects or environmental insults, might lead to disorders of cognitive function. Work in our laboratory has identified a genetic program that is regulated by neuronal activity and controls the experience-dependent shaping of the nervous system. As described in this article, defects in this genetic program underlie disorders of cognition, including autism spectrum disorders.

II. THE MAKING OF A SYNAPSE

During the development of the central nervous system, axons are guided toward their dendritic targets by extracellular guidance cues

The Harvey Lectures, Series 102, pages 73–102
©2010 by John Wiley & Sons, Inc.

(reviewed by Charron and Tessier-Lavigne, 2005). Once the axon has made its way to the target cell with which it will synapse, the initial contact between the axon of the presynaptic cell and the dendrite of the postsynaptic cell is independent of neuronal activity. However, once contact is made, glutamate is released and the development and maturation of the developing synapses occurs in a neuronal activity-dependent manner. This release of neurotransmitter (glutamate at excitatory synapses) results in the depolarization of the plasma membrane and the influx of calcium into the neuron that sets in motion a cascade of events that are critical for processes such as synaptic maturation or elimination. After synapses form early in development, more than half of the synapses in some regions of the brain are eliminated by a process that is calcium- and neuronal activity-dependent (reviewed by Hua and Smith, 2004; Spitzer, 2006).

As excitatory synapses mature, inhibitory synapses also form to balance neuronal excitation, allowing for proper brain function (reviewed by Spitzer, 2006). When the neurotransmitter glutamate is released at an excitatory synapse in response to an environmental stimulus, it binds to N-methyl-D-aspartate (NMDA) and α-amino-3-hydroxy-5-methylisoxazole-4-propionic acid (AMPA) receptors on the postsynaptic cell (Schoepfer et al., 1994; Clements et al., 1998; Ozawa et al., 1998). When this occurs, there is an increased likelihood that that neuron will fire an action potential. When the inhibitory neurotransmitter γ-Aminobotyric acid GABA is released at inhibitory synapses and binds to its receptor, there is a decreased likelihood that that neuron will fire an action potential. The balance between excitation and inhibition is critical for proper brain function, as defects in the maintenance of this balance appear to be the cause of cognitive disorders (Buschges and Manira, 1998; Rubenstein and Merzenich, 2003; Maffei et al., 2004; Cline, 2005). As will be discussed below, our laboratory has recently identified activity-dependent events that function to maintain this balance between excitation and inhibition.

During synaptic activity, membrane depolarization and calcium influx lead to a variety of changes in the cell. Some of these changes occur rapidly and locally such as posttranslational modifications of proteins at the synapse (Pak and Sheng, 2003; DiAntonio and Hicke, 2004; Kim and Sheng, 2004; Tomita et al., 2005; Lin et al., 2006). Other changes in response to membrane depolarization and calcium influx have a longer time course, and they set in motion a cascade of events that lead to the synthesis of new messenger RNAs (mRNAs) and proteins (reviewed in West et al., 2002; Hong et al., 2005). These more long-lasting events can be subdivided into two phases. First, as synapses are forming fairly synchronously in an activity-independent manner during early development, one might imagine that there is enough synaptic drive and glutamate release to lead to depolarization of the plasma membrane, opening of calcium channels, and influx of calcium. Next, this calcium signal is transmitted to the nucleus to initiate new gene transcription. The resulting protein products might then modify the synapse, and thus the biology of the cell. The process of local protein translation is also thought to affect synapse development and may be important for long-lasting changes in synaptic function that are synapse specific (reviewed by Sutton and Schuman, 2006). One idea that is appealing is that the processes of activity-dependent gene transcription and activity-dependent local translation are coupled. Activity-induced mRNAs may be transported to various locations within the cell and then translated locally. Once translated into proteins, the mRNAs may be degraded, thus requiring more gene transcription to replenish the supply of local mRNAs.

The focus of this Harvey Lecture will be our identification and characterization of the neuronal activity-dependent genetic program that regulates synapse development. First, I will describe the experiments that led to the discovery of this genetic program. Next, I will describe the specific components of the program – the genes and proteins involved. Then, I will discuss how this program gets activated at the synapse, both in culture and in vivo in an intact animal. And finally, I will explain how the disruption of this activity-dependent genetic program is responsible for diseases of cognition and may underlie disorders such as autism spectrum disorders.

III. Early Clues of a Genetic Program That Regulates Synapse Development

The first clue for the existence of the neuronal activity-dependent genetic program came from experiments that Dr. Ed Ziff and I did at New York University Medical School in 1984 (Greenberg and Ziff, 1984). We wanted to know the mechanism by which a proliferating cell that is arrested in its cell cycle by growth factor deprivation reenters the cell cycle upon growth factor addition. This was a question that was very

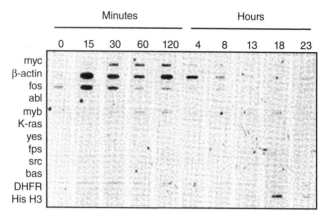

Fig. 4.1. Transcriptional assay of serum-inducible genes. Nuclear run-on assay of transcription of proto-oncogenes in growth-arrested BALB/3T3 cells showing c-fos induction after 15 minutes of serum stimulation. (Adapted from Greenberg and Ziff, 1984, with permission from *Nature*.)

relevant to cancer biology and the basic mechanisms of cell growth. We thought that perhaps the proto-oncogenes that were identified at the time might be activated by growth factors and constitute an RNA synthesis-dependent process that was important for cell cycle progression. In a very primitive gene array experiment, we spotted all the known proto-oncogenes on filter paper and performed a nuclear run-on assay to determine which genes were activated as a cell is stimulated by growth factor addition to progress from G_0 back into the cell cycle. We found that one gene in particular that encoded the proto-oncogene c-Fos was activated rapidly and transiently as the cell marches from G_0 into G_1 and then into DNA synthesis (S) and M phase (Fig. 4.1). This surprising result suggested that growth factors activate genes very quickly, not within hours but within minutes, and we hypothesized that this process might have an important function in cell cycle progression. Subsequently, we and others found that c-*fos* can be activated in response to hormones, growth factors, stress, or any one of a number of environmental stimuli in virtually every cell in the body, both during development and in the mature nervous system. One question that intrigued us was this – can we induce c-*fos* in the brain in response to neuronal activity, and might this give us a handle on the mechanism by which experience leads to changes at synapses that might be important for synaptic development and even human cognitive function? With regard to the experiments of Hubel and Wiesel, we thought that such a finding might present a window into understanding mechanistically how experience shapes the development of the nervous system.

To begin our studies of the role of c-Fos in nervous system development, Ed Ziff, Lloyd Greene, and I took PC12 cells in culture and depolarized them with elevated levels of potassium chloride to open L-type voltage-sensitive calcium channels (L-VSCC) and allow calcium entry into the cell (Greenberg et al., 1986). In response to membrane depolarization, we observed a very rapid activation of c-*fos* transcription literally within a minute of membrane depolarization. This observation was interesting in that it was the first demonstration that a channel in a neural membrane could send a signal from the cell surface to the nucleus to initiate gene transcription. Previously, channels were thought to act locally to affect membrane potential, and to lead to local changes at synapses possibly via the phosphorylation of synaptic proteins. However, the idea that channels could rapidly send a signal to the nucleus was completely surprising and worthy of further study.

We and other investigators soon extended these studies and found that the neuronal activity-dependent activation of gene expression is not specific to PC12 cells. The induction of c-*fos* occurs in cultured neurons in response to glutamate stimulation, and it occurs in the intact brain in response to a variety of physiological stimuli (Morgan et al., 1987; Morgan and Curran, 1991). Around the same time, we and others showed that c-Fos is a transcription factor that, together with c-Jun, regulates the expression of other genes, which in turn trigger cellular responses that depend on the cell-type context and the developmental history of the cell (reviewed in Morgan et al., 1987; Halazonetis et al., 1988; Sheng and Greenberg, 1990; Morgan and Curran, 1991).

The activity-regulated induction of c-*fos* transcription in neurons is part of a pathway that re-sets the cell and allows developing and mature neurons to respond to changes in their environment. c-Fos plays a role in a variety of neuronal functions ranging from neuronal differentiation to activity-dependent adaptive responses in the mature nervous system. In addition, the activation of c-*fos* in a particular region of the brain in response to a physiological stimulus appears to be critical for aspects of the organism's response to the stimulus. For example, the activation of c-*fos* in the suprachiasmatic nucleus (SCN), the region of the brain that is the pacemaker for the generation of a circadian rhythm, is relevant to the behavioral response that one sees during circadian entrainment (Kornhauser et al., 1990). Fear-conditioned stimuli can lead to c-*fos* mRNA induction in the rat amygdala, the region of the brain that mediates fear responses (Campeau et al., 1991). c-*fos* and other immediate early genes are induced in the brain following seizure activity (Morgan et al., 1987). However, the activation of c-*fos* is just the tip of the iceberg. c-*fos* is one member of a large family of activity-regulated genes that are rapidly expressed in response to synaptic activity (reviewed in Hong et al., 2005). Over many years, we and others have characterized this activity-dependent program of gene expression and are trying to understand both how it is activated and what it does.

IV. Approaches to Identifying Components of an Activity-Regulated Genetic Program

To understand how activity-regulated gene expression is regulated, we developed an approach of tracing the pathway in reverse from the nucleus back to the membrane where the gene induction event is initiated. We began by identifying key regulatory elements near the site of initiation of gene transcription. We do this by making mutations in the activity-regulated promoter to define the key regulatory sequences that are important for the activation process. This is a rather difficult task if a particular gene is regulated at multiple levels in that there may be control regions that are many kilobases away, or there may be fine regulation at the level of chromatin structure. Once the DNA regulatory sequences are identified, we determine which transcription factors bind to the elements. Toward this end, we have used gel mobility shift assays to monitor the presence of the transcription factor during its purification, made antibodies to the purified transcription factors, and asked if they are modified and activated in response to neuronal activity by phosphorylation, acetylation, or ubiquitination. If the transcription factors are posttranslationally modified, we then ask what are the signaling transduction pathways that lead to the modification. For example, is there a kinase cascade emanating from the channel where the calcium enters on the cell? Using this approach, we have learned a great deal about the signaling networks that carry signals from synapses to the nucleus to induce the expression of c-*fos* and other activity-regulated genes (reviewed in Shaywitz and Greenberg, 1999; West et al., 2002).

Recently, we have taken a more genome-wide approach to identify the components of the activity-regulated gene program. Hippocampal neurons were grown in culture until the time that synapses are forming and maturing. We then depolarized the neurons to mimic the effects of glutamate release and found that within minutes of membrane depolarization, more than 300 different genes are activated (Lin et al., 2008). These genes have a variety of different functions: many of them encode transcription factors that are activated in every cell of the body, but have specific functions in the nervous system; some are ubiquitin proteosome components; other genes encode signaling molecules; and many are proteins that function specifically at synapses. These approximately 300 activity-regulated genes can be divided into a variety of different groups based on their kinetics of activation (Fig. 4.2). It is worth noting that

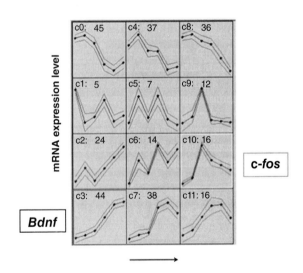

Time after stimulation

Fig. 4.2. Gene expression profiles after stimulation of hippocampal neurons with elevated levels of KCl. Kinetics of induction of genes by 55 mM of KCl in cultured neurons fall into distinct expression profiles. Time after KCl stimulation is along the x-axis; level of mRNA induction along the y-axis. Two genes, *Bdnf* and c-*fos*, are indicated.

expression of the activity-regulated genes is tightly controlled. For example, in the absence of any stimulus, there is no detectable expression of c-*fos*. However, within minutes of release of a transmitter, the c-*fos* gene is activated and expressed at a very high level. In addition, the c-*fos* mRNA turns over very quickly, indicating that there is cellular machinery that shuts off transcription and very rapidly degrades the c-*fos* message. Our laboratory has investigated in detail how c-*fos* and other activity-regulated genes are controlled in an effort to understand this basic cellular process and because defects in the regulation of this genetic program appear to underlie disorders of cognition. If we could identify the components of the c-*fos* regulatory system, we thought we might learn something about normal brain development and what goes wrong in disorders of cognition.

To identify the various components of the protein complexes that regulate activity-dependent gene transcription, we have, in more recent studies, used biochemical approaches to isolate transcription complexes from the brain. We are also using whole-genome RNAi screens in which we set up a reporter assay and ask about each gene in the genome and the ability of its encoded protein to modulate the expression of an activity-regulated reporter gene (Hua and Greenberg, unpublished observations). These new approaches are proving to be very useful for identifying new components of the signaling complexes that control the activity-dependent gene program.

V. COMPONENTS OF AN ACTIVITY-DEPENDENT GENETIC PROGRAM

BDNF is a neurotrophic factor that plays a role in a variety of aspects of nervous system development, including neuronal survival, maturation of inhibitory neurons, and potentiation of synaptic function (Bonni et al., 1999; Huang et al., 1999; Poo, 2001). In addition, *Bdnf* expression is induced by neuronal activity (Tao et al., 1998). BDNF has been implicated in human memory and cognition (Egan et al., 2003), and genetic linkage studies have suggested that BDNF may play a role in schizophrenia and obsessive-compulsive disorder (Krebs et al., 2000). The *Bdnf* gene is very complex, spanning over 40 kilobases of the mammalian genome (Fig. 4.3). It has at least six promoters that give rise to more than 18 different mRNA transcripts in rodents through alternative splicing and differential polyadenylation (Liu et al., 2006; Aid et al., 2007). However,

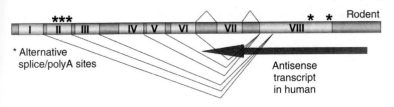

18 possible transcripts

Fig. 4.3. The complex organization of the *Bdnf* gene. The rodent genomic region depicted spans over 40 kilobases and includes eight exons, six possible transcription start sites, nine alternative splice donor sites, and two alternative polyadenylation sites. Because the entire coding region of BDNF is contained in exon VIII, all 18 possible transcripts encode an identical BDNF protein.

each of these *Bdnf* mRNA transcripts gives rise to an identical BDNF protein since the entire coding region is contained within a single exon, exon VIII, that is present in each of the 18 different mRNA transcripts. It is not known why so many transcripts are needed to encode a single protein; however, it is easy to imagine that differential regulation of various promoters and splicing may generate *Bdnf* mRNA transcripts that are temporally or spatially regulated within a neuron.

We have found that *Bdnf* is selectively activated upon calcium influx into neurons. In addition, *Bdnf* can be activated in vivo by a variety of physiological stimuli; for example, visual stimulation early in life will lead to *Bdnf* induction specifically in the visual cortex (Huang et al., 1999). Fear conditioning will activate *Bdnf* in the amygdala (Chhatwal et al., 2006). *Bdnf* is also induced throughout the cortex and the hippocampus during a seizure in an animal (Ernfors et al., 1991). These events can be recapitulated in a tissue culture dish by exposure of cultured neurons to elevated levels of KCl or glutamate receptor agonists (Tao et al., 1998). By quantitative PCR, we have been able to show that promoters I and IV are the most activity-responsive of the *Bdnf* promoters (Tao et al., 1998). Currently, our laboratory is generating mice that are deficient in particular *Bdnf* promoters in an effort to uncover the specific functions of each of the *Bdnf* transcripts.

Using the various approaches mentioned above, we have found that the chromatin surrounding *Bdnf* promoter IV is highly condensed, and that the histones in the chromatin complex are highly methylated (Chen et al., 2003b). This was initially surprising because we expected that since the *Bdnf* gene is rapidly induced within minutes of calcium entry into a neuron, the *Bdnf* promoter IV might not be bound by methylated histones that typically are found associated with genes that are permanently silenced. We have identified several of the transcription factors that are bound to *Bdnf* promoter IV in its repressed state. One of the first transcription factors we identified is a protein called calcium responsive factor (CARF). CARF is a novel transcription factor that appears to confer calcium and neural specificity to *Bdnf* expression (Tao et al., 2002).

Another factor we identified that binds to *Bdnf* promoter IV is a protein called upstream stimulatory factor (USF). USF was one of the first mammalian transcription factors to be identified (Sawadogo and Roeder, 1985), is highly expressed in the nervous system, and appears to bind to the promoters of many activity-regulated genes (Chen et al., 2003a).

A third factor we found to be bound to *Bdnf* promoter IV is the cAMP response element binding protein (CREB). We had previously identified CREB as a key regulator of c-*fos* transcription and demonstrated that CREB is a mediator of calcium-dependent activation of genes in the nervous system (Sheng et al., 1991). Our subsequent work on CREB showed that it is phosphorylated and activated in response to calcium entry into neurons (Sheng et al., 1991; Ginty et al., 1993; Ghosh et al., 1994). When Dr. David Ginty was a postdoctoral fellow in the laboratory, he generated phospho-specific antibodies to a specific phosphorylation site on CREB (Ser133) that since then have been used extensively to study calcium activation of CREB in the intact nervous system (Ginty et al., 1993). Through the use of these antibodies and through many subsequent studies, CREB has been found to play a central role in the development and function of the mature nervous system (reviewed in Shaywitz and Greenberg, 1999). It is possible that many of CREB's functions in the developing nervous system are mediated through its activation of *Bdnf*.

Recently, we made the surprising discovery that methyl CpG-binding protein 2 (MeCP2) is also bound to *Bdnf* promoter IV prior to stimulation (Chen et al., 2003b; Martinowich et al., 2003). MeCP2 binds to methylated cytosines and functions as a long-range transcriptional repressor responsible for gene silencing (Nan et al., 1997; Jones et al., 1998). When MeCP2 binds to methylated DNA, it recruits methyltransferases to methylate nearby CpG's, thereby spreading gene silencing along a region of DNA so that genes in the region where MeCP2 is bound are maintained in an off state. Proteins that bind to methylated DNA, including MeCP2, are thought to be responsible for turning off neural genes in organs other than the brain and turning off nonneural genes in the brain.

We have obtained evidence that MeCP2 may also be part of a repressor complex that ensures that certain neural genes such as *Bdnf* are kept off in the absence of neuronal activity. The finding that MeCP2 is bound to the *Bdnf* promoter in the absence of neuronal activity was surprising since it was not clear why a protein involved in long range and permanent silencing of genes would be bound to the promoter of a gene like *Bdnf* that needs to be quickly activated in response to external stimuli. However, consistent with the hypothesis that MeCP2 is a repressor of *Bdnf*, we observed an increase in the expression of *Bdnf* promoter IV transcripts in MeCP2 knockout mice (Chen et al., 2003b). One possible model that could explain this finding is that upon membrane depolarization and calcium influx into neurons, the MeCP2-dependent repression of *Bdnf* promoter IV is relieved and a protein complex composed of MeCP2, histones, protein deacetylases, and methyltransferases disengages from the promoter. Subsequently, CREB and possibly other positively regulating transcription factors are phosphorylated and activated, the histones are acetylated, and *Bdnf* is expressed transiently at the right time and place in response to neuronal activity. BDNF turned on in this way may then play an important role in synapse development and maturation.

Interestingly, mutations (including deletions and duplications) in *MeCP2* can give rise to a disorder in humans called Rett syndrome (Amir et al., 1999). Rett syndrome is an X-linked disorder that affects approximately 1 in 10,000 females. Girls who have mutations in MeCP2 appear normal at birth and for the first year of life appear to develop normally. At about 1 year of age, they typically learn their first words and they learn to walk. It is precisely at this time during child development that experience leads to a refinement of synaptic connections in the developing brain. In girls with Rett syndrome at approximately 1 year of age, there is a dramatic cognitive regression that can result in severe mental retardation, autism-like behavior, loss of speech, and loss of motor coordination. Recently, mouse models that have altered levels of MeCP2 have been shown to recapitulate the human disorder and provide a means of studying the function of MeCP2 during brain development (Chen et al., 2001;

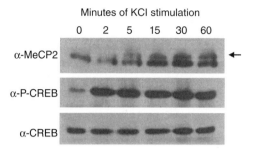

Fig. 4.4.   Phosphorylation of MeCP2 after KCl stimulation. Western blot of cultured neuronal lysates stimulated with 55 mM of KCl for the indicated number of minutes. MeCP2 expression is monitored with a total α-MeCP2 antibody and indicates the presence of a slower migrating species after 5 minutes of KCl stimulation (arrow). Total CREB is used as a loading control for these lysates, while phospho-Ser133 CREB demonstrates the activity-dependent phosphorylation of CREB. (Adapted from Chen et al., 2003a, with permission from *Science*.)

Guy et al., 2001; Shahbazian et al., 2002; Tudor et al., 2002; Dani et al., 2005). It has been found that while MeCP2 is expressed in many cells of the body, the Rett phenotype is specifically neurological. Thus, studies of MeCP2 function in the nervous system may lead to insights into the pathology of Rett syndrome.

We found that MeCP2 is modified upon membrane depolarization, resulting in its slower migration on an SDS polyacrylamide gene and its altered interaction with the *Bdnf* promoter (Chen et al., 2003b). Through a combination of mutagenesis and mass spectrometry, we were able to show that MeCP2 is phosphorylated at serine 421 in response to membrane depolarization in vitro (Fig. 4.4) and physiological stimuli in vivo (Zhou et al., 2006). Using phospho-specific antibodies that we generated that recognize the Ser421 phosphorylated form of MeCP2, we found that while MeCP2 is expressed in many tissues, the phosphorylation of MeCP2 Ser421 appears to occur selectively in the brain. This finding could explain why the loss of MeCP2 in humans and mice has a specific neurological phenotype. It is possible that this activity-dependent function of MeCP2 is layered on top of its function as a long-range silencer of gene expression specifically in the brain as synapses are forming and maturing.

To investigate the relevance of the MeCP2 Ser421 phosphorylation event in the intact nervous system, the MeCP2 Ser421 phospho-specific antibody was used to study MeCP2 phosphorylation in a circadian entrainment protocol (Zhou et al., 2006). In collaboration with Drs. Charles Weitz and Wendy Zhao at Harvard Medical School, mice were entrained to a circadian rhythm so that every 24 hours, they show loco-motor activity on a running wheel. If these mice were then placed in the dark for many days, they exhibited the same locomotor activity approxi-mately every 24 hours, and during a specific interval during the 24-hour circadian cycle. However, when the entrained animals were exposed to light at a time that they perceive as nighttime, the rhythm shifted. When one shifts the circadian rhythm with light, the light stimulates the retinal ganglion neurons to release glutamate onto the neurons of the SCN, the pacemaker for the generation of circadian rhythm; this then leads to the induction of activity-dependent gene expression in the SCN (Kornhauser et al., 1990). In this experimental paradigm, we found that when light stimulation of the retinal ganglia neurons triggers the release glutamate onto the neurons of the SCN, MeCP2 becomes phosphorylated at Ser421 in the SCN as shown using the MeCP2 Ser421 phospho-specific antibody (Zhou et al., 2006). We believe this change in MeCP2 may trigger modi-fications of synaptic function in the SCN and other regions of the brain where neuronal activity induces MeCP2 phosphorylation at Ser421. One way to test this hypothesis is to generate an animal that has replaced Ser421 on MeCP2 with a nonphosphorylatable alanine residue; thus, mice carrying this mutation cannot undergo phosphorylation at this site on MeCP2 in response to physiological stimuli. We have generated these mice and are testing the effects of the Ser421Ala mutation on synaptic development, neural function, and behavior.

As an alternative approach to investigating the importance of MeCP2 Ser421 phosphorylation for synaptic development, we examined the effects of overexpressing wild-type and Ser421Ala mutant MeCP2 protein in hippocampal slices and under these conditions compared their ability to regulate dendrite and dendritic spine development. Examining the effects of overexpressing MeCP2 on synapse development may provide insight into Rett syndrome, given the recent finding that Rett syndrome can arise from a duplication of *MeCP2* as well as from mutations in *MeCP2* that lead to a loss of the protein (del Gaudio et al., 2006). To mimic the effects of duplication of *MeCP2*, biolistics was used to

a

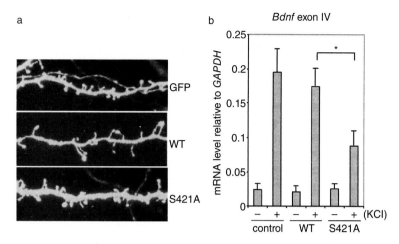

Fig. 4.5.   Effects of MeCP2 phosphorylation at Ser421. (a) Overexpression of wild-type (WT) MeCP2 results in more immature dendritic spines. (b) Loss of phosphorylation of MeCP2 at Ser421 by mutation to Ala leads to a decrease in *Bdnf* mRNA induction by KCl in cultured neurons. (Adapted from Zhou et al. 2006, with permission from *Neuron*.)

introduce extra copies of the gene that encodes MeCP2 into a tissue slice preparation of the mouse hippocampus. We then looked at the effect of overexpressing MeCP2 on synapse development and asked if there is any change in synapse development upon introduction of a Ser421Ala mutated form of MeCP2. For these experiments, a lentivirus vector was used that encodes an RNAi targeting the endogenous *MeCP2* gene as well as an RNAi-resistant flag-tagged version of either the wild-type or Ser-421Ala mutant form of *MeCP2*. We found that in the absence of endogenous MeCP2 (knocked down by RNAi), overexpression of wild-type flag-tagged MeCP2 leads to dendritic spines that are longer and thinner than in the vector control slice (Fig. 4.5a) (Zhou et al., 2006). Long thin spines are thought to be immature spines. It is thus possible that the cognitive impairment seen in Rett syndrome cases caused by overexpression of MeCP2 may be due to immature dendritic spine development, leading to a change in excitatory synaptic function. By contrast, we found that in the absence of endogenous MeCP2, overexpression of the Ser421Ala mutant form of MeCP2 in hippocampal slice preparations leads to spines that have the same length and width as the vector control.

This suggests that dendritic spine development is regulated by the neuronal activity-dependent phosphorylation of MeCP2 Ser421.

We next tested if phosphorylation of MeCP2 at Ser421 affects *Bdnf* gene expression. Again, biolistics was used to introduce into hippocampal neurons a lentivirus vector encoding an RNAi targeting the endogenous MeCP2 as well as a flag-tagged version of either the wild-type or Ser-421Ala mutant forms of MeCP2 that are RNAi resistant. Introduction of the Ser421Ala mutant form of MeCP2, but not wild-type MeCP2, resulted in a significant decrease in *Bdnf* promoter IV-driven mRNA transcripts (Fig. 4.5b). These experiments indicate that phosphorylation of MeCP2 at Ser421 is important for the proper regulation of *Bdnf* promoter IV-dependent transcription.

To understand the mechanism by which calcium influx triggers MeCP2 Ser421 phosphorylation in neurons, we asked which kinase(s) might mediate this event (Zhou et al., 2006). Cortical neurons were incubated with a variety of kinase inhibitors and then exposed to elevated levels of extracellular potassium chloride or NMDA receptor (NMDAR) agonists, and the cells were assessed for the induction of MeCP2 S421 phosphorylation. We found that incubation with KN-93, an inhibitor of CamKII, blocks the activity-regulated phosphorylation of MeCP2 at S421, whereas other pharmacological inhibitors had no effect (Zhou et al., 2006). In addition, phosphorylation of MeCP2 at Ser421 was also blocked by CaMKIIN, a potent and specific endogenous protein inhibitor of CaMKII. By contrast, increasing CaMKII levels by transfecting cultured neurons with a constitutively active mutant form of CaMKII promotes the phosphorylation of MeCP2 at S421 even in the absence of membrane depolarization. Together, these results suggest that CaMKII mediates the phosphorylation of MeCP2 at S421 in response to neuronal activity-induced calcium influx.

We next asked whether MeCP2 is a direct substrate of CaMKII in vitro. Wild-type and Ser421Ala mutant forms of FLAG-tagged MeCP2 were purified from HEK 293T cells and analyzed in an in vitro kinase assay for their ability to be phosphorylated by constitutively active CaMKII. Recombinant CaMKII efficiently phosphorylates wild-type MeCP2 at Ser421 in vitro, but significantly less phosphorylated MeCP2 was observed when Ser421Ala mutant MeCP2 was incubated with CaMKII (Zhou et al., 2006). Taken together, these results indicate that CaMKII mediates the neural-specific phosphorylation of MeCP2 at

Ser421 in response to synaptic activation and subsequent calcium influx. Whether CaMKII directly phosphorylates MeCP2 in vivo or serves as an intermediate kinase in the cascade of events that trigger MeCP2 phosphorylation remains to be determined.

VI. MeCP2 and a Model for Activity-Regulated Gene Expression

These studies have revealed a new function for MeCP2 apart from its role as a long-range silencer of gene expression. Layered on top of, and possibly working together with, its long-range silencer function is MeCP2's ability to regulate specific gene expression in an activity-dependent manner. MeCP2 may be bound to methylated DNA at a locus of control for activity-dependent gene regulation. In the absence of neuronal activity, MeCP2 may repress the expression of the *Bdnf* gene by binding to this locus of control. Neuronal activity then appears to modify MeCP2 in a way that allows for release from this long-range silencing role and the activation of *Bdnf* promoter IV-dependent transcription.

MeCP2 is one component of an intricate signaling network that mediates neurotransmitter-dependent gene activation. When glutamate is released at the developing synapse, it binds to the AMPA and NMDA subtypes of glutamate receptors, leading to membrane depolarization and an influx of calcium typically through the NMDA subtype of glutamate receptors. If there is sufficient membrane depolarization that L-VSCCs are also activated, a calcium signal is sent to the nucleus to activate gene transcription. Both the NMDARs and the L-VSCCs associate with proteins through their cytoplasmic domains that actually function together with calcium as it enters the neuron to convey the calcium signal to the nucleus via particular signaling pathways including the Ras/MAP kinase pathway as well as calmodulin-dependent kinase cascades (reviewed in Dolmetsch, 2003). At the moment the calcium signal first reaches the nucleus, the DNA surrounding genes such as *Bdnf* is methylated and the chromatin is condensed and bound to repressor complexes. However, within minutes of neurotransmitter release at synapses, calcium influx into the postsynaptic cell, and kinase cascade signaling to the nucleus, CREB becomes phosphorylated at a variety of sites (Ser133, 142, and 143), MeCP2 becomes phosphorylated at Ser421, histones become acetylated, and the polymerase II transcription complex is recruited to

Bdnf promoter IV. This sequence of events leads to the activation of *Bdnf* transcription by mechanisms that are not yet completely understood.

A striking feature of the signaling networks that control activity-dependent *Bdnf* transcription is that they are reminiscent of the proto-oncogene oncogene connection that has been seen in proliferating cells and cancer cells in which cell growth and proliferation are deregulated. We have found that the pathways of activity-dependent gene regulation are studded with proteins that are mutated in disorders of cognition (See Fig. 4.8). Mutations in the L-type channel that initiates calcium signaling to the nucleus are implicated in the Timothy syndrome, a disorder that is characterized by heart arrhythmia, syndactyly, and autistic behavior (Splawski et al., 2006). In the mental retardation syndrome Rubinstein–Taybi syndrome, the CREB-binding protein, CBP, is mutated (Petrij et al., 1995). Mutations or polymorphisms in *Bdnf* itself cause memory defects (Egan et al., 2003); and, as discussed here, mutations in *MeCP2* give rise to Rett syndrome. Thus, an understanding of the activity-dependent regulation of *Bdnf* and the other approximately 300 activity-regulated genes has the potential to provide important new insights into the development of human cognition and into how the disruption of this process by mutation of components of the activity-regulated gene networks can lead to cognitive impairment.

VII. The Functions of Two Activity-Regulated Genes: MEF2 and Naps4

In addition to their regulation, we have been investigating the functions of the activity-regulated genes, and toward that end we and others have begun to look at these genes one by one with an emphasis on examining their role in synapse development and maturation (Tao et al., 1998; Fujino et al., 2003; Flavell et al., 2006; Plath et al., 2006; Rial Verde et al., 2006; Shalizi et al., 2006; Shepherd et al., 2006). I describe here our recent findings on several components of the activity-regulated gene program that we have studied in detail. The transcription factors that mediate the activity-dependent gene response are particularly interesting because once a biological function has been identified for the transcription factor, it is possible, using new technologies, to identify each of the gene targets of the transcription factor, and thus define the biological program that mediates the transcription factor's biological effect.

A. MEF2

MEF2(A-D) is a family of four genes expressed in neurons and many other cell types. Certain MEF2 family members are expressed prior to any stimulus and others are activated by neuronal activity (Black and Olson, 1998). For example, MEF2 activity can be regulated by neurotrophins and neuronal activity (Mao et al., 1999; Shalizi et al., 2003; Flavell et al., 2006), and our laboratory has implicated MEF2 in activity-dependent neuronal survival (Mao et al., 1999). We also found that MEF2 family members are highly expressed in the nervous system just at the time of synaptogenesis (Flavell et al., 2006). Using RNAi to knock down MEF2 expression in cultured hippocampal neurons, we detected an increase in the number of excitatory synapses formed on neurons that expressed MEF2 RNAi. In addition, we found by electrophysiological analysis of miniature excitatory postsynaptic currents (mEPSCs), as indicated by an increase in mini frequency, that these extra synapses are functional (Fig. 4.6b). Moreover, when an inducible form of MEF2 was introduced into neurons in which the number of synapses had plateaued and MEF2 activity was constitutively turned on, we found that the number of synapses formed on the MEF2-expressing neuron decreased. These findings suggest that MEF2 may function to restrict excitatory

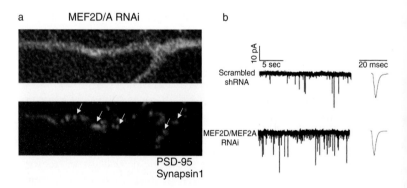

Fig. 4.6.   MEF2 restriction of excitatory synapse number. Increased frequency of spontaneous mini EPSCs in the presence of MEF2 RNAi indicate that the extra excitatory synaptic sites are functional.

synapse number. To investigate how activity alters MEF2 function in the neurons, we have used mass spectrometry to identify activity-dependent modifications of MEF2. We found that in response to neuronal activity and calcium influx, MEF2 is dephosphorylated at multiple sites by a calcium-dependent phosphatase, calcineurin, and that dephosphorylation at these sites is critical for MEF2's ability to restrict synapse number (Flavell et al., 2006). Thus, MEF2 may regulate the activity-dependent synaptic restriction process that is critical to the maturation of neural circuits.

However, several questions remain to be addressed. Does MEF2 function to restrict synapse formation in vivo? What are the targets of MEF2 that mediate its effect on synapse formation, and could MEF2 targets contribute to the elimination of some synapses but not others? One possible scenario is that after synapses form early in development in an activity-independent manner, some synapses may be stimulated more than others. From these active synapses, a calcium signal may be sent to the nucleus leading to the activation of MEF2. Subsequently, the active MEF2 transcription factor may induce the expression of target genes that encode proteins that may function selectively at the synapses that were initially potentiated. Alternatively, activated MEF2 may activate target genes encoding proteins that function to deconstruct synapses that were not initially well activated. In this way, the MEF2 targets may be able to distinguish active synapses from less active synapses, allowing the less active synapses to be eliminated.

To characterize the mechanism by which MEF2 restricts synapse number, we identified MEF2 targets in neurons undergoing synapse development and maturation (Flavell et al., 2008). MEF2 targets were identified with a high level of confidence if they met the following criteria: the gene is an activity-regulated gene that is induced as synapses form and mature; the expression of the gene is significantly reduced in the presence of MEF2 RNAi; the gene is activated by MEF2-VP16, a form of MEF2 that is fused to a strong constitutive transcription activator and is inducible by tamoxifen (Fig. 4.7). Over 100 different target genes satisfy at least two of the three criteria and were validated as MEF2 targets by additional tests. One of these genes encodes SynGAP, a protein first identified as a negative regulator of the small G protein Ras that suppresses Ras activity at synapses (Kim et al., 1998; Oh et al., 2004). MEF2-regulation of SynGAP could explain MEF2's ability to restrict

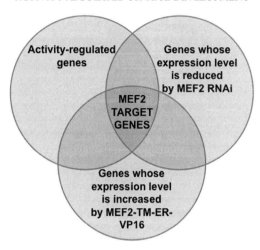

Fig. 4.7.   Strategy of search for MEF2 target genes. Three separate screens were performed to identify MEF2 target genes. The first screen was for activity regulation by elevated levels of KCl. The second screen was for genes whose level of expression was decreased by MEF2 RNAi. The third screen was for genes whose level of expression was increased by an inducible constitutively active form of MEF2. The intersection of these three screens represents the targets of MEF2 that are validated at a high level of confidence.

synapse number in the following way: MEF2 induces the expression of SynGAP, then SynGAP negatively regulates Ras, a mediator of NMDAR signaling, at synapses. This suppression of Ras signaling may contribute to the decrease in synapse number under conditions where MEF2 is active. The second MEF2 target we have identified is Arc, a protein that promotes the internalization of AMPA receptors, a subtype of glutamate receptors that regulate excitatory synaptic transmission. In response to membrane depolarization, MEF2 induces Arc expression leading to AMPA receptors internalization and a loss of functional synapses. Thus, Arc is another MEF2 target that appears to contribute to MEF2's ability to restrict synapse function.

The third MEF2 target we have characterized is a protein called Ube3a, a ubiquitin ligase that is involved in protein degradation. The role that Ube3a plays in the nervous system has not been clear, although Ube3a is mutated in a disorder of cognition called Angelman syndrome characterized by several clinical features including mental retardation, motor dysfunction, and speech impairment. We have found that in the brain, MEF2 promotes the transcription of a subset of *Ube3a* mRNA transcripts, and that the UBE3a protein encoded by these transcripts functions at synapses to restrict synapse number by promoting the degradation of key synaptic proteins (P. Greer and M.E. Greenberg, manuscript in review). Loss of Ube3a recapitulates the MEF2 knockdown phenotype (i.e. there is an increase in synapse number in the absence of Ube3a). It appears that Ube3a is a ubiquitin ligase that regulates the ubiquitination of synaptic proteins, and that this process promotes the restriction or elimination of synapses by a mechanism that remains to be defined.

The identification of these three MEF2 targets – SynGAP, Arc, and Ube3a – has provided new insight into how MEF2 restricts synapse number. We propose that when neurotransmitter is released at a synapse and calcium enters the postsynaptic cell, MEF2 then promotes a program of gene expression that includes the SynGAP, Arc, and Ube3a mRNAs whose protein products function to restrict synapse number.

B. Npas4

Npas4, a member of the bHLH-PAS transcription factor family, has a conserved bHLH-PAS domain at its N-terminus that mediates dimerization with other bHLH-PAS proteins to regulate gene transcription as well as the binding of Npas4 to DNA (Ooe et al., 2004). We observed that *Npas4* expression is transiently induced upon membrane depolarization of cultured cortical and hippocampal neurons (Lin et al., 2008). When neurons are allowed to form synapses in culture and are then exposed to bicuculline, a GABA receptor antagonist that relieves inhibitory inputs so that glutamate is released at excitatory synapses, *Npas4* transcription is induced. Recent studies have shown that transcription of the *Npas4* gene is induced by seizures and ischemia in the intact organism (Flood et al., 2004; Shamloo et al., 2006). Our studies further demonstrated that mice exposed to light for 2–3 hours after 1 week of dark rearing (P21-P28) show a marked increase in the number of neurons in the visual cortex that are immunopositive for Npas4, compared with mice that are dark reared but not exposed to light.

To elucidate the function of Npas4, we used RNAi to knock down Npas4 expression in cultured neurons and assessed the effect on synaptic development (Lin et al., 2008). We found that Npas4 knockdown has no effect on neuronal survival and a very small effect on dendritic growth. In addition, we observed no effect of the loss of Npas4 on excitatory synapse number. However, when we quantified the number of inhibitory synapses in neurons expressing Npas4 RNAi or a control RNAi by measuring the juxtaposition of synaptic markers at the pre- and postsynaptic termini, we detected a very clear reduction in inhibitory synapse number in the presence of Npas4 RNAi. Quantification of this effect shows a clear loss of the inhibitory synapses that form on the dendrites as well as the perisomatic region of the Npas4 RNAi-expressing neuron. Importantly, this effect is recapitulated in an Npas4 knockout mouse. In contrast to the loss of function phenotype, when more copies or the *Npas4* gene are introduced into a neuron, we observed an increase in inhibitory synapses that form on the neuron. One possible explanation for these findings is that when glutamate stimulates calcium entry into neurons and induces *Npas4* transcription, Npas4 turns on a program of gene expression that serves as a beacon to instruct incoming inhibitory presynaptic termini to form or stabilize on the Npas4-expressing cells.

We hypothesize that MEF2 and Npas4 may work together to promote a balance between excitation and inhibition. When excitatory synapses release glutamate, they promote calcium entry into the postsynaptic terminal of the neuron. A signal is then sent to the nucleus, turning on a transcriptional program. MEF2 is activated and restricts excitatory synapse number, thus dampening the excitatory drive. Somewhat later, Npas4 is activated and functions to turn on a program of gene expression that regulates the development of inhibitory synapses on the neuron. An important next step will be to determine the genetic program that is regulated by Npas4. A strategy similar to the one we have developed for identifying MEF2 targets should be useful for identifying Npas4 targets that control the activity-dependent development of inhibitory synapses.

VIII. Concluding Remarks

In summary, our laboratory has uncovered an activity-dependent gene program, one function of which is to control the formation and maturation of synapses and neural circuits in the brain. At a developmental time when synapses are forming and maturing, neuronal activity induces the expression of over 300 different genes in the developing neuron. These genes are activated through an extensive network of signaling that involves

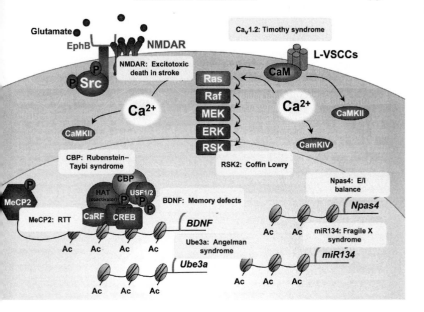

Fig. 4.8.   Calcium signaling and diseases of cognition. Many molecules involved in carrying the calcium signal from the membrane to the nucleus are mutated in humans in disorders of cognition. Defects in Ca$_v$1.2 calcium channel lead to Timothy syndrome (Splawski et al., 2005); NMDAR lead to excitotoxic death in stroke (Mody and MacDonald, 1995); CREB binding protein (CBP) lead to Rubenstein–Taybi syndrome (Petrij et al., 1995); BDNF lead to memory defects (Egan et al., 2003); RSK2 lead to Coffin Lowry (Jacquot et al., 1998); MeCP2 to Rett syndrome (Amir et al., 1999); Ube3a to Angelman syndrome; miR134 appears to regulate LimK mRNA translation through a proposed interaction with Fragile X protein, FMR1, that is mutated in Fragile X syndrome (Schratt et al., 2006); and Npas4 mutation is proposed to lead to defects in the Excitatory/ Inhibitory (E/I) balance. (Lin et al., 2008).

calcium signaling to the nucleus to regulate transcription factor complexes and changes in chromatin that allow genes to be activated. The activated genes play key roles in many aspects of synaptic development, including dendritic growth and synaptic maturation. What is striking about this program is that both the regulators of the genes that we have uncovered in these studies and the genes themselves are mutated in a variety of disorders of cognition (Fig. 4.8). This suggests that intensive study of this genetic program in years to come has the potential to reveal much about the mechanisms of synapse formation and maturation, how synapses operate to lay down thoughts and memories, and other aspects of cognitive function. Understanding how these proteins function will also aid in therapeutic advances to treat individuals with disorders of cognitive function in which mutations cause these pathways to go awry.

Acknowledgments

I thank Dr. Janine Zieg for her extraordinary help in the preparation of this manuscript. I acknowledge the generous support of the F.M. Kirby Foundation to Children's Hospital. This work was supported by the Rett Syndrome Research Foundation, the National Institutes of Health (NS048276), and the Mental Retardation and Developmental Disabilities Research Center Grant (HD018655).

References

Aid, T., Kazantseva, A., Piirsoo, M., Palm, K., and Timmusk, T. 2007. Mouse and rat BDNF gene structure and expression revisited. *J Neurosci Res* **85**:525–535.

Amir, R.E., Van den Veyver, I.B., Wan, M., Tran, C.Q., Francke, U., and Zoghbi, H.Y. 1999. Rett syndrome is caused by mutations in X-linked MECP2, encoding methyl-CpG-binding protein 2. *Nat Genet* **23**: 185–188.

Black, B.L. and Olson, E.N. 1998. Transcriptional control of muscle development by myocyte enhancer factor-2 (MEF2) proteins. *Annu Rev Cell Dev Biol* **14**:167–196.

Bonni, A., Brunet, A., West, A.E., Datta, S.R., Takasu, M.A., and Greenberg, M.E. 1999. Cell survival promoted by the Ras-MAPK signaling pathway by transcription-dependent and -independent mechanisms. *Science* **286**: 1358–1362.

Buschges, A. and Manira, A.E. 1998. Sensory pathways and their modulation in the control of locomotion. *Curr Opin Neurobiol* **8**:733–739.

Campeau, S., Hayward, M.D., Hope, B.T., Rosen, J.B., Nestler, E.J., and Davis, M. 1991. Induction of the c-fos proto-oncogene in rat amygdala during unconditioned and conditioned fear. *Brain Res* **565**:349–352.

Charron, F. and Tessier-Lavigne, M. 2005. Novel brain wiring functions for classical morphogens: a role as graded positional cues in axon guidance. *Development* **132**:2251–2262.

Chen, R.Z., Akbarian, S., Tudor, M., and Jaenisch, R. 2001. Deficiency of methyl-CpG binding protein-2 in CNS neurons results in a Rett-like phenotype in mice. *Nat Genet* **27**:327–331.

Chen, W.G., West, A.E., Tao, X., Corfas, G., Szentirmay, M.N., Sawadogo, M., Vinson, C., and Greenberg, M.E. 2003a. Upstream stimulatory factors are mediators of Ca2+-responsive transcription in neurons. *J Neurosci* **23**:2572–2581.

Chen, W.G., Chang, Q., Lin, Y., Meissner, A., West, A.E., Griffith, E.C., Jaenisch, R., and Greenberg, M.E. 2003b. Derepression of BDNF transcription involves calcium-dependent phosphorylation of MeCP2. *Science* **302**: 885–889.

Chhatwal, J.P., Stanek-Rattiner, L., Davis, M., and Ressler, K.J. 2006. Amygdala BDNF signaling is required for consolidation but not encoding of extinction. *Nat Neurosci* **9**:870–872.

Clements, J.D., Feltz, A., Sahara, Y., and Westbrook, G.L. 1998. Activation kinetics of AMPA receptor channels reveal the number of functional agonist binding sites. *J Neurosci* **18**:119–127.

Cline, H. 2005. Synaptogenesis: a balancing act between excitation and inhibition. *Curr Biol* **15**:R203–R205.

Dani, V.S., Chang, Q., Maffei, A., Turrigiano, G.G., Jaenisch, R., and Nelson, S.B. 2005. Reduced cortical activity due to a shift in the balance between excitation and inhibition in a mouse model of Rett syndrome. *Proc Natl Acad Sci U S A* **102**:12560–12565.

DiAntonio, A. and Hicke, L. 2004. Ubiquitin-dependent regulation of the synapse. *Annu Rev Neurosci* **27**:223–246.

Dolmetsch, R. 2003. Excitation-transcription coupling: signaling by ion channels to the nucleus. *Sci STKE* **2003**:PE4.

Egan, M.F., Kojima, M., Callicott, J.H., Goldberg, T.E., Kolachana, B.S., Bertolino, A., Zaitsev, E., Gold, B., Goldman, D., Dean, M., Lu, B., and Weinberger, D.R. 2003. The BDNF val66met polymorphism affects activity-dependent secretion of BDNF and human memory and hippocampal function. *Cell* **112**:257–269.

Ernfors, P., Bengzon, J., Kokaia, Z., Persson, H., and Lindvall, O. 1991. Increased levels of messenger RNAs for neurotrophic factors in the brain during kindling epileptogenesis. *Neuron* **7**:165–176.

Flavell, S.W., Kim, T.-K., Gray, J.M., Harmin, D.A., Hemberg, M., Hong, E.J., Markenscoff-Papadimitriou, E., Bear, D.M., and Greenberg, M.E. 2008. Genome-wide analysis of MEF2 transcriptional program reveals synaptic target genes and neuronal activity-dependent polyadenylation site selection. *Neuron* **60**:1022–1038.

Flavell, S.W., Cowan, C.W., Kim, T.K., Greer, P.L., Lin, Y., Paradis, S., Griffith, E.C., Hu, L.S., Chen, C., and Greenberg, M.E. 2006. Activity-dependent regulation of MEF2 transcription factors suppresses excitatory synapse number. *Science* **311**:1008–1012.

Flood, W.D., Moyer, R.W., Tsykin, A., Sutherland, G.R., and Koblar, S.A. 2004. Nxf and Fbxo33: novel seizure-responsive genes in mice. *Eur J Neurosci* **20**:1819–1826.

Fujino, T., Lee, W.C., and Nedivi, E. 2003. Regulation of cpg15 by signaling pathways that mediate synaptic plasticity. *Mol Cell Neurosci* **24**:538–554.

del Gaudio, D., Fang, P., Scaglia, F., Ward, P.A., Craigen, W.J., Glaze D.G., Neul, J.L., Patel, A., Lee, J.A., Irons, M., Berry, S.A., Pursley, A.A., Grebe, T.A., Freedenberg, D., Martin, R.A., Hsich, G.E., Khera, J.R., Friedman, N.R., Zoghbi, H.Y., Eng, C.M., Lupski, J.R., Beaudet, A.L., Cheung, S.W., and Roa, B.B. 2006. Increased MECP2 gene copy number as the result of genomic duplication in neurodevelopmentally delayed males. *Genet Med* **8**:784–792.

Ghosh, A., Ginty, D.D., Bading, H., and Greenberg, M.E. 1994. Calcium regulation of gene expression in neuronal cells. *J Neurobiol* **25**:294–303.

Ginty, D.D., Kornhauser, J.M., Thompson, M.A., Bading, H., Mayo, K.E., Takahashi, J.S., and Greenberg, M.E. 1993. Regulation of CREB phosphorylation in the suprachiasmatic nucleus by light and a circadian clock. *Science* **260**:238–241.

Greenberg, M.E. and Ziff, E.B. 1984. Stimulation of 3T3 cells induces transcription of the c-fos proto-oncogene. *Nature* **311**:433–438.

Greenberg, M.E., Ziff, E.B., and Greene, L.A. 1986. Stimulation of neuronal acetylcholine receptors induces rapid gene transcription. *Science* **234**: 80–83.

Guy, J., Hendrich, B., Holmes, M., Martin, J.E., and Bird, A. 2001. A mouse Mecp2-null mutation causes neurological symptoms that mimic Rett syndrome. *Nat Genet* **27**:322–326.

Halazonetis, T.D., Georgopoulos, K., Greenberg, M.E., and Leder, P. 1988. c-Jun dimerizes with itself and with c-Fos, forming complexes of different DNA binding affinities. *Cell* **55**:917–924.

Hong, E.J., West, A.E., and Greenberg, M.E. 2005. Transcriptional control of cognitive development. *Curr Opin Neurobiol* **15**:21–28.

Hua, J.Y. and Smith, S.J. 2004. Neural activity and the dynamics of central nervous system development. *Nat Neurosci* **7**:327–332.

Huang, Z.J., Kirkwood, A., Pizzorusso, T., Porciatti, V., Morales, B., Bear, M.F., Maffei, L., and Tonegawa, S. 1999. BDNF regulates the maturation of inhibition and the critical period of plasticity in mouse visual cortex. *Cell* **98**:739–755.

Jacquot, S., Merienne, K., De Cesare, D., Pannetier, S., Mandel, J.L., Sassone-Corsi, P., and Hanauer, A. 1998. Mutation analysis of the RSK2 gene in Coffin-Lowry patients: extensive allelic heterogeneity and a high rate of de novo mutations. *Am J Hum Genet* **63**:1631–1640.

Jones, P.L., Veenstra, G.J., Wade, P.A., Vermaak, D., Kass, S.U., Landsberger, N., Strouboulis, J., and Wolffe, A.P. 1998. Methylated DNA and MeCP2 recruit histone deacetylase to repress transcription. *Nat Genet* **19**:187–191.

Kim, E. and Sheng, M. 2004. PDZ domain proteins of synapses. *Nat Rev Neurosci* **5**:771–781.

Kim, J.H., Liao, D., Lau, L.F., and Huganir, R.L. 1998. SynGAP: a synaptic RasGAP that associates with the PSD-95/SAP90 protein family. *Neuron* **20**:683–691.

Kornhauser, J.M., Nelson, D.E., Mayo, K.E., and Takahashi, J.S. 1990. Photic and circadian regulation of c-fos gene expression in the hamster suprachiasmatic nucleus. *Neuron* **5**:127–134.

Krebs, M.O., Guillin, O., Bourdell, M.C., Schwartz, J.C., Olie, J.P., Poirier, M.F., and Sokoloff, P. 2000. Brain derived neurotrophic factor (BDNF) gene variants association with age at onset and therapeutic response in schizophrenia. *Mol Psychiatry* **5**:558–562.

Lin, Y., Jover-Mengual, T., Wong, J., Bennett, M.V., and Zukin, R.S. 2006. PSD-95 and PKC converge in regulating NMDA receptor trafficking and gating. *Proc Natl Acad Sci U S A* **103**:19902–19907.

Lin, Y., Bloodgood, B.L., Hauser, J.L., Lapan, A.D., Koon, A.C., Kim, T.K., Hu, L.S., Malik, A.N., and Greenberg, M.E. 2008. Activity-dependent regulation of inhibitory synapse development by Npas4. *Nature* **455**:1198–1204.

Liu, Q.R., Lu, L., Zhu, X.G., Gong, J.P., Shaham, Y., and Uhl, G.R. 2006. Rodent BDNF genes, novel promoters, novel splice variants, and regulation by cocaine. *Brain Res* **1067**:1–12.

Maffei, A., Nelson, S.B., and Turrigiano, G.G. 2004. Selective reconfiguration of layer 4 visual cortical circuitry by visual deprivation. *Nat Neurosci* **7**:1353–1359.

Mao, Z., Bonni, A., Xia, F., Nadal-Vicens, M., and Greenberg, M.E. 1999. Neuronal activity-dependent cell survival mediated by transcription factor MEF2. *Science* **286**:785–790.

Martinowich, K., Hattori, D., Wu, H., Fouse, S., He, F., Hu, Y., Fan, G., and Sun, Y.E. 2003. DNA methylation-related chromatin remodeling in activity-dependent BDNF gene regulation. *Science* **302**:890–893.

Mody, I. and MacDonald, J.F. 1995. NMDA receptor-dependent excitotoxicity: the role of intracellular Ca2+ release. *Trends Pharmacol Sci* **16**:356–359.

Morgan, J.I. and Curran, T. 1991. Stimulus-transcription coupling in the nervous system: involvement of the inducible proto-oncogenes fos and jun. *Annu Rev Neurosci* **14**:421–451.

Morgan, J.I., Cohen, D.R., Hempstead, J.L., and Curran, T. 1987. Mapping patterns of c-fos expression in the central nervous system after seizure. *Science* **237**:192–197.

Nan, X., Campoy, F.J., and Bird, A. 1997. MeCP2 is a transcriptional repressor with abundant binding sites in genomic chromatin. *Cell* **88**:471–481.

Oh, J.S., Manzerra, P., and Kennedy, M.B. 2004. Regulation of the neuron-specific Ras GTPase-activating protein, synGAP, by Ca2+/calmodulin-dependent protein kinase II. *J Biol Chem* **279**:17980–17988.

Ooe, N., Saito, K., Mikami, N., Nakatuka, I., and Kaneko, H. 2004. Identification of a novel basic helix-loop-helix-PAS factor, NXF, reveals a Sim2 competitive, positive regulatory role in dendritic-cytoskeleton modulator drebrin gene expression. *Mol Cell Biol* **24**:608–616.

Ozawa, S., Kamiya, H., and Tsuzuki, K. 1998. Glutamate receptors in the mammalian central nervous system. *Prog Neurobiol* **54**:581–618.

Pak, D.T. and Sheng, M. 2003. Targeted protein degradation and synapse remodeling by an inducible protein kinase. *Science* **302**:1368–1373.

Petrij, F., Giles, R.H., Dauwerse, H.G., Saris, J.J., Hennekam, R.C., Masuno, M., Tommerup, N., van Ommen, G.J., Goodman, R.H., Peters, D.J., and Breuning, M.H. 1995. Rubinstein-Taybi syndrome caused by mutations in the transcriptional co-activator CBP. *Nature* **376**:348–351.

Plath, N., Ohana, O., Dammermann, B., Errington, M.L., Schmitz, D., Gross, C., Mao, X., Engelsberg, A., Mahlke, C., Welzl, H., Kobalz, U., Stawraka-kis, A., Fernandez, E., Waltereit, R., Bick-Sander, A., Therstappen, E., Cooke, S.F., Blanquet, V., Wurst, W., Salmen, B., Bosl, M.R., Lipp, H.P., Grant, S.G., Bliss, T.V., Wolfer, D.P., and Kuhl, D. 2006. Arc/Arg3.1 is essential for the consolidation of synaptic plasticity and memories. *Neuron* **52**:437–444.

Poo, M.M. 2001. Neurotrophins as synaptic modulators. *Nat Rev Neurosci* **2**:24–32.

Rial Verde, E.M., Lee-Osbourne, J., Worley, P.F., Malinow, R., and Cline, H.T. 2006. Increased expression of the immediate-early gene arc/arg3.1 reduces AMPA receptor-mediated synaptic transmission. *Neuron* **52**:461–474.

Rubenstein, J.L. and Merzenich, M.M. 2003. Model of autism: increased ratio of excitation/inhibition in key neural systems. *Genes Brain Behav* **2**:255–267.

Sawadogo, M. and Roeder, R.G. 1985. Interaction of a gene-specific transcription factor with the adenovirus major late promoter upstream of the TATA box region. *Cell* **43**:165–175.

Schoepfer, R., Monyer, H., Sommer, B., Wisden, W., Sprengel, R., Kuner, T., Lomeli, H., Herb, A., Kohler, M., Burnashev, N., Gunther, W., Ruppersberg, P., and Seeburg, P. 1994. Molecular biology of glutamate receptors. *Prog Neurobiol* **42**:353–357.

Schratt, G.M., Tuebing, F., Nigh, E.A., Kane, C.G., Sabatini, M.E., Kiebler, M., and Greenberg, M.E. 2006. A brain-specific microRNA regulates dendritic spine development. *Nature* **439**:283–289.

Shahbazian, M., Young, J., Yuva-Paylor, L., Spencer, C., Antalffy, B., Noebels, J., Armstrong, D., Paylor, R., and Zoghbi, H. 2002. Mice with truncated MeCP2 recapitulate many Rett syndrome features and display hyperacetylation of histone H3. *Neuron* **35**:243–254.

Shalizi, A., Lehtinen, M., Gaudilliere, B., Donovan, N., Han, J., Konishi, Y., and Bonni, A. 2003. Characterization of a neurotrophin signaling mechanism that mediates neuron survival in a temporally specific pattern. *J Neurosci* **23**:7326–7336.

Shalizi, A., Gaudilliere, B., Yuan, Z., Stegmuller, J., Shirogane, T., Ge, Q., Tan, Y., Schulman, B., Harper, J.W., and Bonni, A. 2006. A calcium-regulated MEF2 sumoylation switch controls postsynaptic differentiation. *Science* **311**:1012–1017.

Shamloo, M., Soriano, L., von Schack, D., Rickhag, M., Chin, D.J., Gonzalez-Zulueta, M., Gido, G., Urfer, R., Wieloch, T., and Nikolich, K. 2006. Npas4, a novel helix-loop-helix PAS domain protein, is regulated in response to cerebral ischemia. *Eur J Neurosci* **24**:2705–2720.

Shaywitz, A.J. and Greenberg, M.E. 1999. CREB: a stimulus-induced transcription factor activated by a diverse array of extracellular signals. *Annu Rev Biochem* **68**:821–861.

Sheng, M. and Greenberg, M.E. 1990. The regulation and function of c-fos and other immediate early genes in the nervous system. *Neuron* **4**:477–485.

Sheng, M., Thompson, M.A., and Greenberg, M.E. 1991. CREB: a Ca(2+)-regulated transcription factor phosphorylated by calmodulin-dependent kinases. *Science* **252**:1427–1430.

Shepherd, J.D., Rumbaugh, G., Wu, J., Chowdhury, S., Plath, N., Kuhl, D., Huganir, R.L., and Worley, P.F. 2006. Arc/Arg3.1 mediates homeostatic synaptic scaling of AMPA receptors. *Neuron* **52**:475–484.

Spitzer, N.C. 2006. Electrical activity in early neuronal development. *Nature* **444**:707–712.

Splawski, I., Timothy, K.W., Decher, N., Kumar, P., Sachse, F.B., Beggs, A.H., Sanguinetti, M.C., and Keating, M.T. 2005. Severe arrhythmia disorder caused by cardiac L-type calcium channel mutations. *Proc Natl Acad Sci U S A* **102**:8089–8096; discussion 8086–8088.

Splawski, I., Yoo, D.S., Stotz, S.C., Cherry, A., Clapham, D.E., and Keating, M.T. 2006. CACNA1H mutations in autism spectrum disorders. *J Biol Chem* **281**:22085–22091.

Sutton, M.A. and Schuman, E.M. 2006. Dendritic protein synthesis, synaptic plasticity, and memory. *Cell* **127**:49–58.

Tao, X., Finkbeiner, S., Arnold, D.B., Shaywitz, A.J., and Greenberg, M.E. 1998. Ca2+ influx regulates BDNF transcription by a CREB family transcription factor-dependent mechanism. *Neuron* **20**:709–726.

Tao, X., West, A.E., Chen, W.G., Corfas, G., and Greenberg, M.E. 2002. A calcium-responsive transcription factor, CaRF, that regulates neuronal activity-dependent expression of BDNF. *Neuron* **33**:383–395.

Tomita, S., Stein, V., Stocker, T.J., Nicoll, R.A., and Bredt, D.S. 2005. Bidirectional synaptic plasticity regulated by phosphorylation of stargazin-like TARPs. *Neuron* **45**:269–277.

Tudor, M., Akbarian, S., Chen, R.Z., and Jaenisch, R. 2002. Transcriptional profiling of a mouse model for Rett syndrome reveals subtle transcriptional changes in the brain. *Proc Natl Acad Sci U S A* **99**:15536–15541.

West, A.E., Griffith, E.C., and Greenberg, M.E. 2002. Regulation of transcription factors by neuronal activity. *Nat Rev Neurosci* **3**:921–931.

Wiesel, T.N. and Hubel, D.H. 1963. Effects of visual deprivation on morphology and physiology of cells in the cats lateral geniculate body. *J Neurophysiol* **26**:978–993.

Zhou, Z., Hong, E.J., Cohen, S., Zhao, W.N., Ho, H.Y., Schmidt, L., Chen, W.G., Lin, Y., Savner, E., Griffith, E.C., Hu, L., Steen, J.A., Weitz, C.J., and Greenberg, M.E. 2006. Brain-specific phosphorylation of MeCP2 regulates activity-dependent Bdnf transcription, dendritic growth, and spine maturation. *Neuron* **52**:255–269.

CILIA AND HEDGEHOG SIGNALING IN THE MOUSE EMBRYO

KATHRYN V. ANDERSON

Developmental Biology Program, Sloan-Kettering Institute, New York, New York

I. Introduction

Although we humans have a vested interest in understanding human biology, our understanding of mammalian development has lagged far behind our understanding of development in *Drosophila* and *Caenorhabditis elegans*, in large part due to the superior genetic tools available in those invertebrates. However, the status of mouse genetics has improved dramatically in recent years. During the last 15 years, many mutations that affect mouse development have been generated by homologous recombination in mouse embryonic stem cells, and analysis of these mutations has provided fundamental insights into the roles of genes, proteins and genetic pathways that establish the body plan of the mouse embryo. However, despite the intense efforts of many mouse biologists, the phenotypes of mutations in fewer than 20% of all mouse genes have been analyzed, and in most cases, the phenotype has been analyzed in only a single tissue. It seems, therefore, that we have only scratched the surface of a complete description of the genetic control of mammalian development. However, the mouse genome sequence provides an additional powerful resource for mouse genetics. In our lab, we have used phenotype-based genetic screens to identify mutations that disrupt mammalian development and taken advantage of the genome sequence to identify the genes responsible for the mutant phenotypes. Among the genes identified in this approach are the ones that have central, but mammalian-specific, roles in development. One striking example of the success of this phenotype-based genetic screen is the discovery of the essential role for an organelle, the cilium, in mammalian Hedgehog (Hh) signaling.

The Harvey Lectures, Series 102, pages 103–116
©2010 by John Wiley & Sons, Inc.

II. Phenotype-Based Gene Discovery in the Mouse

The approach that has been most successful in the discovery of genes and genetic pathways that regulate specific aspects of biology in the standard model organisms – yeast, *Drosophila*, *C. elegans* – is the phenotype-based genetic screen. In this approach, large numbers of mutations are generated at random, and genes of interest are identified because they have specific effects on the biological process under study. Phenotype-based screens depend on the availability of an efficient mutagen, the ability to breed a relatively large number of animals that are homozygous for newly induced mutations, and easy assays for the phenotype of interest. It has been known for more than two decades that ethylnitrosourea (ENU) is a very potent mutagen in the mouse (Hitotsumachi et al., 1985). This mutagen is efficient enough that screening the progeny of ~700 F1 animals is sufficient to mutate every gene in the genome on an average of one time. The morphology of the midgestation in embryo is rich enough that many mutations that affect embryonic development can be identified by simple visual inspection of the embryos. The well-annotated genome sequence of the mouse makes it straightforward to go from a mutation that causes an interesting phenotype to the identification of the DNA sequence change responsible for that phenotype.

Our lab has carried out a small-scale, but long-term, screen to identify new genes important for mammalian development based on the morphology of midgestation embryos that may be homozygous for new ENU-induced mutations (Fig. 5.1; Kasarskis et al., 1998; García-García et al., 2005). We have identified dozens of mutations that produce clear morphological changes in the embryo and have successfully used genome sequence information to identify the genes responsible for the most interesting phenotypes.

Neural tissue represents a large fraction of the midgestation (e9.5) embryo. It is therefore not surprising that we have identified a relatively large number of mutations that change the shape of the brain. It was, however, surprising, that we identified a set of genes that affected brain patterning that all affect an overlooked organelle, the primary cilium. Here we describe how the genetic screen led to the discovery that the primary cilium is essential for dorsal-ventral patterning of cell types within the developing neural tube.

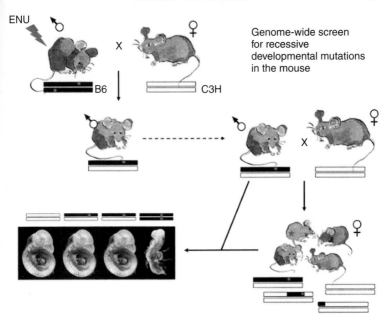

Fig. 5.1.   The crossing scheme for identifying recessive ethylnitrosourea (ENU)-induced mutations that affect morphology of the midgestation embryo. Male mice from the C57Bl6 (B6) strain are treated with ENU and mated with females from a different strain, C3H. Their F1 male progeny, which are heterozygous for approximately 30 new gene-inactivating mutations (star), are mated to C3H females. The second-generation female progeny are mated back to their father, and embryos are examined at midgestation for abnormal morphology. The gene responsible for the phenotype is identified based on the genetic linkage of the phenotype to DNA polymorphisms characteristic of the B6 strain. (Drawing courtesy of M.J. García-García.)

III. Hedgehog Signaling and Dorsal-Ventral Patterning of the Mouse Spinal Cord

Based on chick embryological studies, it has long been known that the dorsal-ventral pattern of cell types in neural tissue is initiated by signals from adjacent, nonneural tissues. Cell types in the dorsal half of the neural tube are organized by signals from the surface ectoderm. These dorsalizing signals are likely to be both Wnts and Bmps, although genetic analysis of these signals has been complicated by early lethality and overlapping function of these gene families. Cell types in the ventral half of the spinal cord are determined by signals from the notochord, and analysis of mouse mutants has made it clear that ventral signal is Sonic Hedgehog (Shh): *Shh* mutants lack six different ventral neural cell types (Chiang et al., 1996).

Hh signaling is crucial for many aspects of normal development in the embryo, and inappropriate signaling can lead to tumor development (Hooper and Scott, 2005). In the mammalian embryo, decreased Hh leads to birth defects that include holoprosencephaly, the loss of ventral midline of brain and face. Excess Hh signaling causes different types of birth defects, including polydactyly. Postnatally, inappropriate Hh signals cause tumors (Rubin and de Sauvage, 2006). Indeed, all cases of the most common human tumor, the basal cell carcinoma, are associated with inappropriate activation of Hh signaling in the skin. In addition, the rare childhood tumors medulloblastoma and rhabdomyosarcoma are frequently caused by activated Hh signaling. Thus, an in-depth understanding of the mechanisms that regulate mammalian Hh signaling is of direct importance for human health.

The core Hh signaling pathway was defined in genetic studies in Drosophila (Hooper and Scott, 2005). Hh, the secreted ligand, activates Patched (Ptch), a transmembrane receptor. Ptch negatively regulates another transmembrane protein, Smoothened (Smo), through a nonstoichiometric mechanism that may be mediated by a small molecule. When Hh binds Ptch, Ptch no longer inhibits the activity of Smo, and Smo, together with other interacting proteins, lead to activation of the Ci transcription factor that implements Hh signals.

Targeted mutations in the mouse have shown that the core Hh pathway is conserved in the mouse (Hooper and Scott, 2005). There are some simple differences between the Drosophila and mammalian Hh pathways: there are three Hh ligands in mammals (of the three, it is Shh that appears to regulate neural patterning) and there are three Ci homologues, Gli1, Gli2 and Gli3. However, there are additional, more fundamental differences between the Drosophila and vertebrate Hh pathways (Huangfu and Anderson, 2006). For example, the structure of the Smo protein is quite different in Drosophila and vertebrates. Activation of Ci in Drosophila depends on the formation of a complex on the C-terminal cytoplasmic domain of Smo; this domain of vertebrate Smo is truncated and lacks motifs that are believed to be essential for binding the Gli-containing complex. To date, the mechanisms that couple Smo to Gli regulation in mammals are not known.

IV. Novel Mutations That Disrupt Hedgehog Signaling

In our screen, we identified nine mutations that change the morphology of the brain and alter the pattern of cell types in the ventral neural tube (Garcia-Garcia et al., 2005). Two of these mutations affect conserved components of the Hh pathway: Smo and Disp1 (which is required for the release of Hh from Hh-producing cells) (Caspary et al., 2002). None of the other mutations mapped to the genomic positions that correspond to characterized Hh pathways components, which suggested that significant aspects of mammalian Hh signaling had not been characterized.

Mutations in four different genes affect a common cellular process with no previous association with Hh signaling: intraflagellar transport (IFT) (Huangfu et al., 2003). For example, *wimple* mutants (named for the shape of their heads, which is reminiscent of a medieval headdress) were identified in the screen because they showed two characteristic patterning phenotypes: randomization of left-right asymmetry (as shown in the direction of heart looping) and a distinctively shaped open brain (exencephaly). In contrast to the simple failure to fuse the two edge of the neural plate, which leads to an open neural tube that has a trough on the ventral midline, *wim* embryos had a ridge of neural tissue that suggested the absence of ventral neural cell types. Indeed, analysis of cell types in the *wim* neural tube showed that the embryos lack a suite of Hh-dependent cell types: Shh is expressed normally in the notochord, but the *wim* neural tube lacks the floor plate, V3 interneurons and most motor neurons, cell types that depend on high and intermediate levels of Hh signaling. The *Ptch* gene, which encodes the Hh receptor, is a direct transcriptional target of the pathway in both Drosophila and vertebrates. In the wild-type neural tube, *Ptch* is expressed in a ventral-to-dorsal gradient that reflects the ventral-to-dorsal gradient of Hh activity. In *wim* embryos, only low levels of *Ptch* are expressed in the neural tube, confirming that Hh signaling is indeed disrupted in the mutant embryos.

V. IFT is Required for Hedgehog Signaling

Standard positional cloning methods were used to identify the gene responsible for the *wim* phenotype (Huangfu et al., 2003). The gene was mapped to a 500 kb region on mouse chromosome 5, and systematic sequencing of the genes in the interval led to the identification of a single missense mutation in the interval that disrupted the *Ift172* gene. IFT genes were discovered in the single-celled alga *Chlamydomonas*, where it was shown that IFT is required for the formation and maintenance of flagella, long cellular extensions built on a microtubule scaffold (Rosenbaum and Witman, 2002). *Chlamydomonas* IFT proteins form large complexes that transport cargo to the flagellar tip, release their cargo at the tip and are reorganized to form a retrograde transport complex that recycles flagellar components back to the cell body. Although the only flagella in mammals are sperm tails, cilia have the same structure as flagella and also depend on IFT. The best-studied cilia in the mouse embryo are the long cilia on the embryonic node that are required for the initiation of left-right asymmetry. *wim* mutants lack all cilia on the node, which confirmed that the gene disrupted in *wim* mutants was *Ift172* and explained the loss of left-right asymmetry in the mutant embryos.

In addition to *wim/Ift172*, we found that three other mutations that disrupt cilia formation also disrupt Hh signaling (Huangfu et al., 2003; Huangfu and Anderson, 2005). We identified a partial loss-of-function allele of *polaris* (*Ift88*) in our screen based on its disrupted neural patterning. A null allele of *polaris* blocks formation of all cilia (Murcia et al., 2000), and we found that the null allele causes the same loss of ventral neural cell types seen in *wim* embryos. Kinesin-II is the motor that powers anterograde IFT (Cole et al., 1998). Null alleles of *Kif3a* lack cilia and disrupt left-right asymmetry (Marszalek et al., 1999; Takeda et al., 1999), and we found that a null allele of *Kif3a* lacks ventral neural cell types; later, we identified an allele of *Kif3a* in the genetic screen. Cytoplasmic dynein 2 is the motor that powers retrograde IFT (Pazour et al., 1999), and we identified two alleles of the gene encoding the heavy chain of this dynein, *Dync2h1*; these mutants also lack ventral neural cell types. The dynein mutants have cilia, but these cilia have bulges along their length that are consistent with a disruption of retrograde trafficking within the cilium. The *Ift88*, *Kif3a* and *Dync2h1* mutants all fail to express normal levels of *Ptch* RNA in the ventral neural tube, demonstrating that Shh signaling is disrupted in these mutants. The common phenotypes of mutations in these four genes argue strongly that IFT is required for Shh signaling.

Double mutant analysis demonstrated that all four of these IFT proteins act at the core of the Shh signal transduction pathway. For example, Ptch is a negative regulator of the pathway and the pathway is activated to high levels in mouse embryos that lack *Ptch*, so that all neural cells

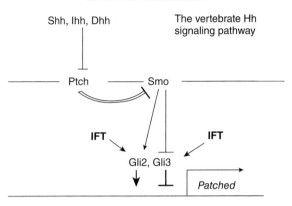

Fig. 5.2. The vertebrate Hedgehog (Hh) signaling pathway. The core of the Hh pathway is evolutionarily conserved: Hh ligands bind to Patched and relieve the repression of Patched on Smoothened (Smo). Activation of Smo promotes formation of the transcriptional activator form of Gli2 and blocks the formation of the Gli3 transcriptional repressor. Intraflagellar transport (IFT) proteins are required for the activity of the pathway in the mouse, but not in Drosophila. IFT proteins have a dual activity: they are required for the formation of both Gli2 activator and Gli3 repressor. Ihh: Indian Hedgehog. Dhh: Desert Hedgehog.

acquire a ventral identity. Double mutants that lack both *wim* and *Ptch* are indistinguishable from *wim* embryos: like *wim*, they lack ventral neural cell types. Because the pathway is not activated by the removal of Ptch when Ift172 is not present, Ift172 must act at a step downstream of Ptch in the pathway. Similar double mutant analysis showed that all four of the Ift genes described here act downstream of Smo but upstream of the Gli proteins (Fig. 5.2; Huangfu et al., 2003; Huangfu and Anderson, 2005).

VI. IFT Mutants Block Both Ligand-Regulated Transcriptional Activation and Repression

Despite the position of the IFT proteins at the heart of the Shh signal transduction pathway, the phenotype caused by loss of the IFT proteins is not identical to the phenotype of mutants that lack Shh. The IFT mutants have a slightly milder phenotype than Shh mutants; for example, the IFT mutants have V1 and V2 interneurons, which are induced by low levels of Hh activity. We found that this was due to the complexity of the Hh pathway. Hh-dependent patterning of the mouse neural tube relies on two transcription factors, Gli2 and Gli3. In the absence of Hh ligand, Gli3 is proteolytically processed into a lower-molecular-weight transcriptional repressor that keeps the pathway off in the absence of ligand. In the presence of Hh, Gli3 processing is blocked and Gli2 is converted into an activator that promotes the transcription of Hh target genes. In the IFT mutants, Gli3 repressor is not made efficiently; as a consequence, Hh target genes are expressed at low levels in the absence of ligand. Loss of IFT proteins prevents both types of responses to Hh ligands: Gli activator is not made and Gli repressor activity is not modified. As a result, Hh target genes are expressed at a low, ligand-independent level.

Many tissues depend on Hh signals for normal development, but some rely primarily on Gli activator-dependent transcriptional induction and others depend on the regulation of Gli repressor production. Specification of the most ventral cell types in the neural tube depends on the production of Gli activator. In contrast, the number of digits in the limbs is controlled by Gli repressor levels and in the absence of Gli3, limbs have six or seven digits rather than five. Those IFT mutants that survive long enough to develop digits show polydactyly, which reflects the loss of Gli3 repressor in that tissue. Thus, the phenotypes of IFT mutants vary between tissues: in tissues that depend on Hh-dependent transcriptional activation IFT phenotypes resemble Shh mutants, and IFT mutants resemble Gli3 mutants in tissues that depend on Gli repressor activity.

VII. Cilia are Required for Hedgehog Signaling in the Mouse, But Not in Drosophila

Given the intensity with which Hh signaling has been studied in Drosophila, it was surprising that the role of IFT genes in Hh signaling was not discovered in flies. IFT genes are present in the Drosophila genomes, and mutations in Drosophila IFT genes have been characterized, but those mutations do not affect Hh signaling. Drosophila IFT mutants are viable and fertile and show behavioral deficits due to disruption of cilia on ciliated sensory neurons (Han et al., 2003). In Drosophila and other invertebrates, cilia are found only on sensory neurons. In contrast, nonmotile primary cilia are found on most vertebrate cells. This evolutionary difference raised the possibility that IFT proteins are required for mouse Hh signaling because mouse Hh signaling depends on the presence of an organelle, the cilium.

A single nonmotile primary cilium is present on many cells in the mouse embryo; in particular, the cells of the developing neural tube that respond to Hh have relatively long cilia that project into the lumen of the neural tube. As first shown by Corbit et al. (2005) and Haycraft et al. (2005), the conserved proteins that mediate Hh signaling are enriched in cilia. Corbit et al. (2005) showed that Smo moves from a cytoplasmic compartment into cilia in response to treatment with Shh. Even mouse embryo fibroblasts in culture have cilia when grown to confluency. In fibroblasts derived from Ptch mutant embryos, the Hh pathway is activated, as can be assayed by Gli-luciferase reporter genes. While Smo is cytoplasmic in untreated wild-type mouse embryo fibroblasts (MEFs), Smo is present in the cilia of Ptch mutant MEFs. These protein localization studies strongly support the hypothesis that cilia are required for mammalian Hh signaling.

One could still argue that the enrichment of Smo, Gli proteins, Sufu and Ptch in cilia is not causal, and that IFT proteins are required for transport between nonciliary cytoplasmic compartments. However, a growing number of mutations that affect cilia structure but do not affect the IFT proteins themselves have been identified, and in every case, these mutations alter Hh signaling. For example, mutations in the mouse homologue of the human disease gene Ofd1, which encodes a basal body protein, prevent formation of cilia and block all signaling through the Hh pathway (Ferrante et al., 2006). Slowly, a compelling case has been built that mammalian Hh signaling depends on the presence of the cilium (Fig. 5.3).

VIII. Disruption of the Ciliary Axoneme Alters Hedgehog Activity

Among the mutants we identified, some mutants have more subtle disruptions of cilia and cause interesting alterations in Hh signaling. The *hennin* (*hnn*) mutation, like the other mutations described thus far, was identified in our screen based on striking changes in the shape of the brain that were accompanied by randomization of the left-right polarity of heart looping (Caspary et al., 2007). Unlike the IFT mutants, however,

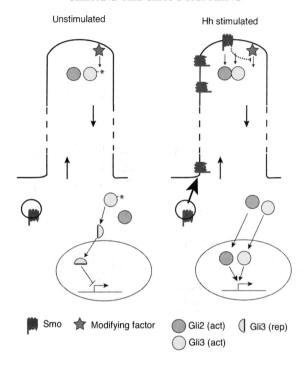

Fig. 5.3. Mammalian Hedgehog (Hh) signal transduction takes place in cilia. In the absence of ligand, cilia are required for the production of the Gli3 repressor that keeps the pathway off in the absence of ligand. In the absence of ligand, Gli proteins are present at cilia tips, but Smoothened (Smo) is not present in cilia. In response to Hh ligand, Smo moves into cilia and leads to changes in Gli proteins that promote the formation of Gli activator and block the formation of Gli3 repressor. (Modified from Eggenschwiler and Anderson, 2007.)

hnn mutants do not show a simple loss of ventral neural cell types. Instead, *hnn* mutants lack the floor plate, the cell type that requires the highest level of Shh signaling, but have many more motor neurons than wild-type embryos. Motor neurons are specified by intermediate levels of Hh signaling. Thus, *hnn* mutants appear to have intermediate (but not high or low) levels of activity of the Hh pathway in most cells of the neural tube. In confirmation of this interpretation, *Ptch* is expressed at intermediate levels throughout much of the *hnn* neural tube.

One possible explanation for the *hennin* phenotype was that *hnn* affected the spreading of extracellular Shh ligand: perhaps in the absence of Hnn, Shh ligand spreads further so that it does not achieve a high ventral concentration but reaches more cells. To test that hypothesis, we analyzed *hnn Shh* double mutants: neural patterning in these embryos was identical to that in *hnn* single mutants. This ruled out the possibility that Hnn regulated Shh diffusion and showed that, instead, the Hnn acts downstream of Shh.

Positional cloning showed that the *hnn* mutation disrupts a small GTPase of the Arl subfamily of Ras-like GTPases, Arl13b (Caspary et al., 2007). Antibodies were raised against Arl13b, and immunostaining showed that Arl13b is highly enriched in the ciliary axoneme. Arl13b is required for normal cilia structure: cilia are about half as long in the node of *hnn* mutants as in wild type. Ultrastructural analysis showed that the *hnn* cilia have an additional defect. The ciliary axoneme is normally composed on a ring of nine microtubule doublets. In *hnn* mutants, one of the microtubules (the so-called B tubule) is open in most doublets. As other mutants with half-length cilia have normal Hh signaling, it is the change in the structure of the axoneme that correlates best with the disruption of Hh signaling.

Analysis of Gli3 processing and of double mutants that lack both Arl13b and components of the core Hh pathway led to the model that Arl13b required both to tether Gli2 in cilia and to promote the transcriptional activator function of Gli2, with little or no effect on Gli3 repressor. Combined with the altered structure of the *hnn* axoneme, the data suggest the hypothesis that Gli2 (or a regulator of Gli2) may associate specifically with the nontubulin component of the ciliary axoneme and that this association is required for proper regulation of Gli2 activation. Thus, changes in the structure of the cilium can have complex effects on Hh signaling.

IX. CILIA IN THE MOUSE EMBRYO ARE COMPLEX MACHINES THAT TRANSDUCE HEDGEHOG SIGNALS

Although cilia are found on nearly all cells in the mouse embryo, the only obvious developmental defects in embryos that lack cilia can be accounted for by disruption in Hh signaling. Loss of canonical or non-canonical Wnt signaling, Fgf receptor signaling or TGFβ-family signaling all cause profound disruption in the morphology of embryos, but the phenotypes of cilia-less mutants do not overlap with those phenotypes. Mutants that lack cilia die before e12.5 due to disruption of Hh signaling, so these strong mutants cannot be analyzed for defects in signaling pathways that act later in development. Nevertheless, it appears that cilia in the early mouse embryo are complex machines whose function is dedicated to the response to Hh ligands.

ACKNOWLEDGMENTS

The work described here was carried out by the students and postdoctoral fellows in the lab, including Andrew Kasarskis, Danwei Huangfu, Andrew Rakeman, Tamara Caspary, María J. García-García and Jonathan Eggenschwiler, and was, in part, a collaboration with Aimin Liu and Lee Niswander.

REFERENCES

Caspary, T., García-García, M.J., Huangfu, D., Eggenschwiler, J.T., Wyler, M.R., Rakeman, A.S., Alcorn, H.L., and Anderson, K.V. 2002. Mouse dispatched homolog1 is required for long-range, but not juxtacrine, Hh signaling. *Curr Biol* **12**(18):1628–1632.

Caspary, T., Larkins, C.E., and Anderson, K.V. 2007. The graded response to Sonic Hedgehog depends on cilia architecture. *Dev Cell* **12**(5):767–778.

Chiang, C., Litingtung, Y., Lee, E., Young, K.E., Corden, J.L., Westphal H., and Beachy, P.A. 1996. Cyclopia and defective axial patterning in mice lacking Sonic Hedgehog gene function. *Nature* **383**:407–413.

Cole, D.G., Diener, D.R., Himelblau, A.L., Beech, P.L., Fuster, J.C., and Rosenbaum, J.L. 1998. Chlamydomonas kinesin-II-dependent intraflagellar transport (IFT): IFT particles contain proteins required for ciliary assembly in *Caenorhabditis elegans* sensory neurons. *J Cell Biol* **141**:993–1008.

Corbit, K.C., Aanstad, P., Singla, V., Norman, A.R., Stainier, D.Y., and Reiter, J.F. 2005. Vertebrate smoothened functions at the primary cilium. *Nature* **437**:1018–1021.

Eggenschwiler, J.T., Anderson, K.V. 2007. Cilia and developmental signaling. *Annu Rev Cell Dev Biol* **23**:345–373.

Ferrante, M.I., Zullo, A., Barra, A., Bimonte, S., Messaddeq, N., et al. 2006. Oral-facial-digital type I protein is required for primary cilia formation and left-right axis specification. *Nat Genet* **38**:112–117.

García-García, M.J., Eggenschwiler, J.T., Caspary, T., Alcorn, H.L., Wyler, M.R., Huangfu, D., Rakeman, A.S., Lee, J.D., Feinberg, E.H., Timmer, J.R., and Anderson, K.V. 2005. Analysis of mouse embryonic patterning and morphogenesis by forward genetics. *Proc Natl Acad Sci U S A* **102**:5913–5919.

Han, Y.G., Kwok, B.H., and Kernan, M.J. 2003. Intraflagellar transport is required in Drosophila to differentiate sensory cilia but not sperm. *Curr Biol* **13**:1679–1686.

Haycraft, C.J., Banizs, B., Aydin-Son, Y., Zhang, Q., Michaud, E.J., and Yoder, B.K. 2005. Gli2 and Gli3 localize to cilia and require the intraflagellar transport protein polaris for processing and function. *PLoS Genet* **1**:e53.

Hitotsumachi, S., Carpenter, D.A., and Russell, W.L. 1985. Dose-repetition increases the mutagenic effectiveness of N-ethyl-N-nitrosourea in mouse spermatogonia. *Proc Natl Acad Sci U S A* **82**:6619–6621.

Hooper, J.E. and Scott, M.P. 2005. Communicating with Hedgehogs. *Nat Rev Mol Cell Biol* **6**:306–317.

Huangfu, D. and Anderson, K.V. 2005. Cilia and Hedgehog responsiveness in the mouse. *Proc Natl Acad Sci USA* **102**:11325–11330.

Huangfu, D. and Anderson, K.V. 2006. Signaling from Smo to Ci/Gli: conservation and divergence of Hedgehog pathways from Drosophila to vertebrates. *Development* **133**:3–14.

Huangfu, D., Liu, A., Rakeman, A.S., Murcia, N.S., Niswander, L., and Anderson, K.V. 2003. Hedgehog signalling in the mouse requires intraflagellar transport proteins. *Nature* **426**:83–87.

Kasarskis, A., Manova, K., and Anderson, K.V. 1998. A phenotype-based screen for embryonic lethal mutations in the mouse. *Proc Natl Acad Sci U S A* **95**:7485–7490.

Marszalek, J.R., Ruiz-Lozano, P., Roberts, E., Chien, K.R., and Goldstein, L.S. 1999. Situs inversus and embryonic ciliary morphogenesis defects in mouse mutants lacking the KIF3A subunit of kinesin-II. *Proc Natl Acad Sci USA* **96**(9):5043–5048.

Murcia, N.S., Richards, W.G., Yoder, B.K., Mucenski, M.L., Dunlap, J.R., and Woychik, R.P. 2000. The Oak Ridge Polycystic Kidney (orpk) disease gene is required for left-right axis determination. *Development* **127**:2347–2355.

Pazour, G.J., Dickert, B.L., and Witman, G.B. 1999. The DHC1b (DHC2) isoform of cytoplasmic dynein is required for flagellar assembly. *J Cell Biol* **144**:473–481.

Rosenbaum, J.L. and Witman, G.B. 2002. Intraflagellar transport. *Nat Rev Mol Cell Biol* **3**:813–825.

Rubin, L.L. and de Sauvage, F.J. 2006. Targeting the Hedgehog pathway in cancer. *Nat Rev Drug Discov* **5**:1026–1033.

Takeda, S., Yonekawa, Y., Tanaka, Y., Okada, Y., Nonaka, S., and Hirokawa, N. 1999. Left-right asymmetry and kinesin superfamily protein KIF3A: new insights in determination of laterality and mesoderm induction by kif3A-/- mice analysis. *J Cell Biol* **145**(4):825–836.

DERIVATION OF ADULT STEM CELLS DURING EMBRYOGENESIS

LEONARD I. ZON

Children's Hospital Boston, Howard Hughes Medical Institute, Boston, Massachusetts

I. Introduction

During embryogenesis, organ development is guided by transcription factors that induce genetic programs specific to each organ. My work has focused on this cell fate determination. In organs such as the heart, blood, lungs, and intestine, stem cells are deposited and differentiated when needed. My laboratory has developed a better understanding of the factors that regulate such stem cell populations.

II. Initiation Rights

In the late 1980s, a number of labs had begun evaluating transcription factors that regulate the activation of the globin genes (Wood, 1983; Orkin, 1990). As the field was progressing, it became evident that GATA sites in individual globin promoters were critical for function. A number of other erythroid-specific genes were similarly being studied for activation of transcription. Surprisingly, the genes encoding globins, heme biosynthetic proteins, and red cell membrane proteins were regulated by GATA sites (Manovani et al., 1987; deBoer et al., 1988; Evans et al., 1988; Galson and Housman, 1988; Gumucio et al., 1988; Wall et al., 1988; Knezetic and Felsenfeld, 1989; Mignotte et al., 1989; Perkins et al., 1989; Plumb et al., 1989). A single DNA-binding protein had the ability to transactivate a number of erythroid genes, and this factor was the master regulator of the erythroid program. The race was on to try to identify the transcription factor.

The Harvey Lectures, Series 102, pages 117–132
©2010 by John Wiley & Sons, Inc.

The typical method at the time to identify transcription factors involved purification strategies, as well as screening phage libraries for the ability of transcription factors to bind labeled double-stranded DNA. The purification strategy was laborious. In the case of GATA factors, the strategy for phage cloning simply did not work. I was a postdoctoral fellow in Stuart Orkin's laboratory and developed an expression cloning scheme for the isolation of this important factor. I constructed pools of cDNAs, transfected them into COS cells and then made nuclear extracts (Orkin, 1990). Gel mobility shifts were done for each nuclear extract, and ultimately, our group identified the gene that bound to GATA as a novel zinc finger transcription factor. It was subsequently named GATA-1.

GATA-1 was the founding member of a family of transcription factors that regulates blood development and also endothelial, kidney, heart, and gut development (Burch, 2005; Cantor, 2005). There are now at least six GATA factor family members in the human genome. Once the cDNA was available, gene expression studies were done by Northern blot analysis of a variety of human and mouse leukemia lines. We found that the transcription factor was expressed in low levels in early multipotential cells, but was expressed at very high levels in erythroid cells and mega-karyocytes (Martin et al., 1990). This finding raised a central question in my mind. Here we had found the erythroid regulator, and yet the ery-throid regulator was expressed in an erythroid-specific fashion. As I started my own laboratory, I decided to work on how the induction of this red cell transcription factor occurs during embryogenesis. I believed that ubiquitous developmental factors would be responsible for a signaling pathway that turned on GATA-1.

III. Model Systems

A. Xenopus

In the early days of my laboratory, I spent time evaluating a number of model systems that would be used to find factors upstream of GATA-1. My initial foray into the mouse was very disappointing. At the time, I wanted to determine when GATA-1 was first expressed in an embryo. My studies led to the dissection of 7.5-day mouse embryos in which the embryos are extremely tiny, and GATA-1 was expressed at 7.5 days. The inability to get large numbers of embryos at the early time point was very frustrating.

The cloning of GATA-1 had been a major focus in a number of laboratories. The chicken GATA-1 clone was isolated by Todd Evans in Gary Felsenfeld's lab and demonstrated some interesting structural differences (Evans and Felsenfeld, 1989). As I was thinking about projects, I had cloned the frog GATA-1 for defining further information about the structure of GATA-1 (Zon et al., 1991). At a gathering of scientists, I ran into Gerry Thompson from Doug Melton's laboratory and told him that I had isolated the frog GATA-1 cDNA clone. He felt that the development of the red blood cell series could be answered best in an externally fertilized animal such as *Xenopus*. *Xenopus* females can produce thousands of eggs in an individual clutch. He recommended that I talk with Doug Melton. Doug was very enthusiastic about my ideas to define the factors that regulate ventral mesoderm and blood formation. This led to a short sabbatical in Doug's laboratory in which I was teamed with Gerry Thompson and Ali Hermanti-Brevanlou to teach me developmental biology. At the time, Doug Melton's lab was one of the premier labs to study developmental biology, and I was very fortunate to be under the tutelage of these two excellent scientists. I already had my own laboratory with postdoctoral fellows and was actually learning *Xenopus* embryology as a faculty member, even though I had never been trained as a postdoctoral fellow on that topic. I have always felt that the best training allows one to learn how to learn, rather than focusing on a given set of techniques.

B. *The Early Days of* Xenopus *Blood Formation*

There were a number of descriptive papers about how blood was formed on the ventral side of the *Xenopus* embryo, but little was known regarding the factors that regulate ventral mesoderm. We developed an expression cloning scheme, very much akin to the GATA-1 cloning, to find such factors. At the time, Richard Harland had done some expression cloning in *Xenopus* embryos to find *Spemann* organizer genes (Smith and Harland, 1991). A factor, wnt8, seemed to be one of the factors regulating dorsal mesoderm formation. In my lab, we injected pools of cDNAs and found a pool that ventralized the embryo (Fig. 6.1). Upon sib selection, this was found to be mix1 (Mead et al., 1996). We next positioned mix1 in a pathway responding to BMP signaling as ventral mesoderm forms.

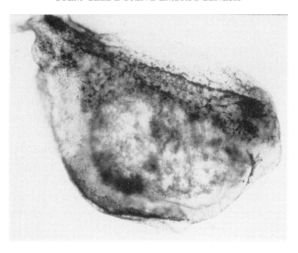

Fig. 6.1.   Expression of mix1 led to a ventralized embryo with no head, and hemoglobin expression throughout the animal (brown).

The role of mix1 in hematopoiesis has recently been studied in the mouse (Robb et al., 2000; Hart et al., 2002; Guo et al., 2002; Mohn et al., 2003; Glaser et al., 2006; Willey et al., 2006). As I was doing these studies, I realized that most of our discoveries were based on overexpression and the characterization of dominant-negative mutants. The lack of genetic backup in the *Xenopus* system made me nervous. I searched for another system, and the zebrafish would occupy most of my laboratory for the next 10 years.

C. The Zebrafish

I had heard about Christian Nusslein-Volhard's entry into the zebrafish field as a genetic system to try to prove the developmental principles she had found in Drosophila applied to vertebrates (Mullins and Nusslein-Volhard, 1993). With a Nobel Prize winner entering into the field and other efforts from Mark Fishman's and Wolfgang Dreiver's labs on large-scale screens affecting organ development, it seemed the right time to enter into a new field. The new zebrafish model might allow me to develop into a leadership role within that field, whereas the *Xenopus* community had many years of leaders. Also, ventral mesoderm appeared to take a backseat.

In 1992, I went to a Hemoglobin Switching Meeting and presented our work on the *Xenopus* factors that induce ventral mesoderm. After my lecture, Frank Grosfeld commented to me that he felt the work was top-notch, but he wondered about genetics. He also had heard about the zebrafish. I explained to him that I was thinking about moving in that direction. He then spent an hour defining what he did in the mouse system, allowing the system to take him where the biology was interesting. I returned the next day to receive a call from Bill Detrich, a professor at Northeastern University, who works on Antarctic fish (Zhao et al., 1998). Bill wanted to understand how cold adaptation of transcription could occur in his ice fish model. I informed him that I was not interested in studying the ice fish, but was thinking about studying the zebrafish. He elected to sign on for a sabbatical to start in July 1993. The next day, Carl Fulwiler from Wally Gilbert's lab informed me that he had a blood-less zebrafish mutant that came about in a screen (Liao et al., 2002), and he wondered if I would like to work on that mutant. Within a week, a number of events led me to enter into the zebrafish field.

The zebrafish is a wonderful system for studying organ development. The embryos are completely transparent, and each mother has about 300 babies per week. The genetic and genomic resources have been developed, and this allows positional cloning to occur of mutant genes in a fairly rapid fashion. At the time when I jumped into the zebrafish field, the genomic resources were really not developed. I organized the field to develop a platform for technology that would help positional cloning. This led to a conversation with NIH Institute Director Harold Varmus; I hoped to develop a genome project for the zebrafish. In an important meeting for the field, I presented to the NIH Institute directors and found them very receptive to starting a genome project. This work continues as the Trans-NIH Zebrafish Initiative (http://www.nih.gov/science/models/zebrafish/).

D. Mutant Fish

At the time of my developing the genome resources, we only had one mutant. In 1995, I heard that Christian Nusslein-Volhard's lab had iso-lated many mutant zebrafish. I contacted her to find out when the blood mutants would be available. She told me that nobody in her lab actually cared about blood and we were welcome to the mutants. I sent a postdoctoral fellow, Dave Ransom, and a graduate student, Alison Brownlie, to Germany for 6 months to work on the mutants there and to bring them back to Children's Hospital. By the end, we had 26 complementation groups of mutants that affected hematopoiesis (Ransom et al., 1996). The screen was to visually look at blood cells circulating and to evaluate mutants that appeared anemic. Our mutants were named after wines because there are red and white blood cells, and red and white wines. This collection of mutants would provide a wealth of knowledge for the entire field.

E. Highlights of Our Zebrafish Work

There were a number of critics of zebrafish hematopoiesis research in the early days. First, would the zebrafish form blood equivalently or using the same genetic pathways as humans? I believed that this would be the case, but others were not so sure. When Alison Brownlie cloned the *sauternes* mutant, she found it to encode the ALAS2 gene (Brownlie et al., 1998). This became the first animal model of congenital sideroblastic anemia, a rare disease of the erythroid lineage. It also became the first positionally cloned mutant gene for a disease in the zebrafish system. This gene established that hematopoiesis is to be conserved throughout all vertebrate development. ALAS2 had already been known to be the cause of congenital sideroblastic anemia in humans. Unfortunately, this gene was not novel, and the effort to positionally clone it was significant. The critics felt that the system would only produce known genes.

In our next mutant gene cloning study, Adriana Donovan isolated a transporter that brought iron from the yolk into the embryo (Donovan et al., 2000). This transporter that we called ferroportin was subsequently found in humans to be involved in the placental iron transporters as pregnant women take iron to allow the baby to be iron replete, and subsequently found to be that the intestinal iron transporter delivers iron from the diet. Later, humans with iron metabolism issues were found that were mutated in ferroportin (Njajou et al., 2001), thus making this zebrafish mutant the first time a new zebrafish mutant developed a new human disease. We had then isolated a novel gene and demonstrated it to be a human disease gene. More recently, we have found two genes called mitoferrin (Shaw et al., 2006) and grx5 (Wingert et al., 2005) as the *frescati* and *shiraz* mutants. These proved to be defective in humans with iron homeostasis problems.

Still, others wished that we would find the factor that control GATA1 during development. In 1993, Alan Davidson isolated the kuggelig mutant, which encoded CDX4 (Davidson et al., 2003; Davidson and Zon, 2006). This demonstrated a CDX-HOX pathway that regulates the body plan to segregate some mesoderm associated with blood. Finally, a factor had been isolated that appeared upstream of GATA1. CDX4 is also known to be a direct wnt target, and we are searching for the factors that regulate its activity.

F. Stem Cell Transplantation in the Zebrafish

Another central question was whether the zebrafish would have hematopoietic stem cells. To answer this, it would require a marrow transplant. In the zebrafish case, the adult site of hematopoiesis is the kidney, so we have done "kidney marrow transplants." A number of people in the lab had tried to do this experiment and had failed miserably. This was mostly because it was always a side project and not a focus of these individuals. I had also written to E. Donald Thomas (the Nobel Prize winner for bone marrow transplantation) to get his advice and he told me that in the 1950s he had tried to do this experiment already. He thought it was "a good project for two high school students." Meanwhile, I was ready to spend a significant amount of our lab's energy on transplantation. David Traver came to the lab after he had done his PhD with Irv Weissman on marrow transplants in the mouse. He knew that the zebrafish system could be adapted. By taking GFP+ (green fluorescent protein) bone marrow and placing it into an irradiated adult or into a mutant that lacked blood stem cells, he was able to reconstitute the immune system with green blood cells (Traver et al., 2003, 2004). These animals could be secondarily transplanted, thereby demonstrating that blood stem cells do exist in the zebrafish system. This allows us a great tool to understand how stem cells are regulated and to evaluate mutants for their ability to improve stem cell homeostasis.

In more recent work, we have begun to evaluate adult stem cell homing. We have developed a transparent adult zebrafish using a combination of pigment mutants. Individual cells can be placed inside the heart, and cells migrate to hematopoietic sites using green fluorescent protein. Such an approach will be invaluable for the field in the future, and we will really have an excellent tool not available in other systems.

G. Recent Work on Stem Cells

We have focused on the derivation of adult stem cells during embryogenesis. There are multiple waves of hematopoietic development during embryogenesis (Galloway and Zon, 2003). The first wave arises on the yolk sac and its predominant use is to make red blood cells for the purpose of carrying oxygen during hypoxic periods of embryogenesis. A second wave derives in the aorta in a region known as the aorta-gonads-mesenephrous region or the AGM. Later, hematopoietic cells in the fetal liver contain stem cell activity, and ultimately, the bone marrow is the adult site of hematopoiesis in the human.

We investigated the aorta in the zebrafish by staining embryos using whole-mount in situ hybridization for runx1 and c-myb, two markers of hematopoietic stem cells (Burns et al., 2002). Runx1 is the product of the AML1 leukemia gene, and c-myb is a proto-oncogene. Both delineate hematopoietic stem cells in a variety of organisms, including *Xenopus*, chicken, mouse, and human, and have been shown to be in the ventral wall of the aorta. We demonstrated that these genes in the zebrafish are similarly expressed. Thus, 300 million years of evolution has conserved the onset of adult hematopoiesis with hematopoietic stem cells in the aorta.

The derivation of this region driving embryogenesis has been studied in a variety of animals including *Xenopus*, chicken, mouse, and human. In several cases, grafting studies have been done. For instance, in the frog embryo, it is possible to make diploid/triploid tissue transplants and follow the contribution of the aorta to adult hematopoiesis (Kau and Turpen, 1983; Turpen and Smith, 1985). It is also possible to use quail-chick chimeras to follow these cells (Dieterlen-Lievre, 1974). In the chick, there are two aortae that are paired and then fused at midline, and the ventral wall becomes hematopoietic. The endoderm expresses an inducing signal for this region before it is formed; also, areas of the somitic mesoderm have growth factors that can alter the fate of these cells (Pardanaud et al., 1996; Dieterlen-Lievre and Cumano, 1998). In the zebrafish, we found that the dorsal part of the dorsal aorta expresses a transcription factor called TBX20, whereas the ventral wall expresses runx1 and c-myb.

At the time we initiated our studies, very little was known about the molecular pathways regulating derivation of this area. We first studied mindbomb, a mutant that lacks notch signaling (Burns et al., 2005). Notch is a receptor on cells that is cleaved in the cytoplasmic domain and becomes a transcription function. Mindbomb is the E3 ubiquitin ligase for delta, a notch ligand. In the cell containing delta, the recycling of delta is required for continuous notch signaling. In the mindbomb mutant, there was a lack of runx1 and c-myb staining. Furthermore, we found that runx1 RNA overexpression could rescue the mindbomb mutant. Using a heat-shock inducible activated notch line, we demonstrated that notch activation led to the entire aorta becoming hematopoietic (both dorsal and ventral). Also, the vein underneath the aorta induced runx1 and c-myb expression. This effect was proven to be a fate change since BRDU labeling and phosphoH3 staining did not reveal any cell cycle abnormality. The notch-runx pathway is an important regulator of definitive hematopoiesis.

We began to examine the role of notch signaling in adult fish. A clinical trial was done in the zebrafish in which animals were sublethally irradiated to reduce their stem cells and over a 30-day recovery period were monitored to evaluate whether activation of the notch pathway would improve recovery. An increased percentage of precursors, myeloid and lymphoid cells was found earlier in the recovery period. This demonstrated that notch signaling had a role in stem cell homeostasis, particularly during stress-activated conditions. In an effort to find other factors that regulate AGM stem cell production, we have undertaken a genetic screen. We have 10 mutants that affect this particular process. These will provide ample opportunities to understand interesting genes affecting stem cell production.

IV. A Chemical That Promotes Self-Renewal of Blood Stem Cells

Another popular technique in the zebrafish system is to utilize chemical genetics as a method to probe pathways. The zebrafish is a wonderful system in which a large number of embryos can be obtained and the size of the embryos is relatively small and can fit in multiwell plates. Two types of chemical screens can be done: forward chemical genetic screens in which the chemicals can be found that affect organ development directly, and also screens for chemicals that suppress a mutant phenotype. One such chemical was persynthemide, a chemical found in our laboratory to suppress a cell cycle mutant in the zebrafish associated with cancer (Stern et al., 2005).

We undertook a screen for small molecules that amplify hematopoietic stem cells (North et al., 2007). For this assay, we made use of wild-type zebrafish and produced over 8000 embryos a week. Individual embryos were placed in wells, and chemicals were aliquoted to those wells. An in situ hybridization with a combined probe for runx and c-myb was done. We evaluated the number of hematopoietic stem cells in the aorta. The screen was done using a library of chemicals of known action. There were 2500 such chemicals in the library and about a third of the library included FDA-approved drugs. The advantage of these chemicals was that a mechanism was known and if we found a particular activity, there was the potential to go to the clinic.

Of the 2500 chemicals that were screened, we found roughly 35 that increased hematopoietic stem cells and about 47 that reduced hemato-poietic stem cells. This provides a wealth of interesting chemicals for modulating stem cell activity. Surprisingly, a number of the chemicals in the library were prostaglandin precursors. We found that inhibitors of prostaglandin synthesis led to a reduction of hematopoietic stem cells in the aorta, whereas chemicals that were prostaglandin precursors appeared to increase hematopoietic stem cells.

We investigated whether individual prostaglandins, such as prostaglan-din E2 (PGE2), I2, F2, and D2, had activity of inducing hematopoietic stem cells. We also found that PGE2 was the most abundant prostaglan-din in the zebrafish embryo and likely to account for the predominant activity. Only PGE2 and prostaglandin I2 were able to increase hemato-poietic stem cells. The induction by PGE2 was not as strong as some of the prostaglandin precursors found in our screen. Since the half-life of PGE2 in circulation is 1 minute, we found a stable derivative of PGE2 that was very active. A stabilized version of PGE2, 16,16-dimethyl PGE2 (dmPGE2), had a substantial activity inducing hematopoietic stem cells (Fig. 6.2). This chemical could be given to individual embryos, and the entire aorta was hematopoietic based on runx1 and c-myb staining. We also evaluated inhibitors of COX1 and COX2 to determine if they would reduce hematopoietic stem cells based on confocal imaging. One of the transgenic lines that we used for this purpose is a double transgenic in which the c-myb BAC transgene has been engineered with GFP at the ATG. Thus, stem cells would turn green. This is mated to another line with the LMO2 promoter driving DsRed in which the blood and blood vessel system is red. Thus, cells that have both colors are likely to be stem

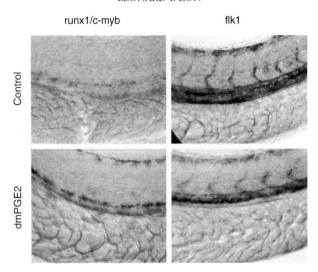

Fig. 6.2. Note the increase in runx1/c-myb expression induced by 16,16-dimethyl PGE2, whereas flk1 expression is normal.

cells. As shown in Figure 6.2, there is an increase in hematopoietic stem cells.

We next evaluated whether PGE2 had the ability to increase hematopoietic stem cells in the irradiation assay described above for notch. We soaked the fish in water containing PGE2 after sublethal irradiation and demonstrated, similarly, that there was an improved recovery. This demonstrated that the chemical we had found in the prostaglandin pathway was active both during embryogenesis and adulthood. We demonstrated that addition of PGE2 to mouse embryoid bodies that are derived from embryonic stem cells leads to an increase in hematopoietic stem cell progenitors. This activity spurred us to try the chemical in adult hematopoietic stem cell homeostasis.

In an effort to evaluate adult hematopoiesis, we removed bone marrow from the mouse and added the dimethyl PGE2 to the marrow for 2 hours and then did a CFU-S (colony forming unit-spleen) assay. CFU-S measures a very early multipotential progenitor or short-term hematopoietic stem cell. PGE2 addition led to a threefold increase in the number of stem cells. To determine whether prostaglandin signaling acted in a cell

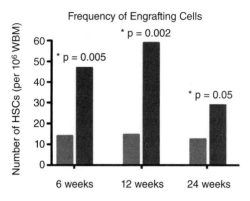

Fig. 6.3. Frequency of engrafting cells over time in a limit dilution competitive repopulation. The red bars were controls, and the blue bars were treated with 16,16-dimethyl PGE2.

autonomous way with regard to stem cells, we purified a stem cell fraction using the c-kit+ sca+ lin− cells. Again, addition of PGE2 led to an increase in CFU-S activity. Addition of indomethacin suppressed CFU-S activity, demonstrating requirement of the process of prostaglandin formation to actually engraft. We also demonstrated in a 5-fluoro recovery assay that PGE2 accelerated recovery similar to our zebrafish work and that specific COX1 and COX2 inhibitors reduced that activity. Lastly, using a competitive transplant limit dilution assay, we were able to demonstrate that there was a threefold increase in stem cells engrafting over a 6-month period (Fig. 6.3). Thus, dmPGE2 is the first small molecule discovered that amplifies stem cells.

The activity of this chemical is remarkable and is similar to the activity of some other peptide growth factors. For instance, wnt3 activation of hematopoietic stem cells ex vivo led to a threefold increase in hematopoietic engraftment in a competitive transplant assay. In addition, a treatment with angiopoietin-like proteins also had a similar increase in activity.

A. Potential Clinical Trials

Over the past 10 years, cord blood transplantation has proven to be a very successful means of treating a variety of genetic diseases as well as cancer. These transplants work very well in children; however, in adults, there is a limit dose of stem cells within individual cords and transplantation often leads to a failure of engraftment due to insufficient stem cells. Thus, agents that increase hematopoietic stem cells in cord blood transplants are valuable. Cord blood has the ability to evade the immune system and is becoming very popular in adult transplants. Recently, two cords have been used from unrelated donors. This is an interesting practice and raises some questions about immune stability. It is clear that two cords appear to be better than one.

We envision a process in which we would use dmPGE2 to stimulate hematopoiesis ex vivo from the cord blood and the relative threefold induced engraftment in the mouse system may confer such an advantage in the human that this becomes clinically applicable. For such a trial, we plan to add PGE2 to a single cord and to compete it in vivo against an untreated cord. In this experiment, similar to the mouse, a higher chimeric engraftment rate should occur for the cord treated with PGE2.

The availability of the zebrafish for understanding diseases, as well as for developing new therapies, positions it at a sweet spot for vertebrate biology. The studies available in the fish interface nicely with mouse and human studies since the gene set has been evolutionarily conserved.

References

deBoer, E., Antoniou, M., Mignotte, V., Wall, L., and Grosveld, F. 1988. The human beta-globin promoter; nuclear protein factors and erythroid specific induction of transcription. *EMBO J* 7:4203–4212.

Brownlie, A., Donovan, A., Pratt, S.J., et al. 1998. Positional cloning of the zebrafish sauternes gene: a model for congenital sideroblastic anaemia. *Nat Genet* **20**:244–250.

Burch, J.B. 2005. Regulation of GATA gene expression during vertebrate development. *Semin Cell Dev Biol* **16**:71–81.

Burns, C.E., DeBlasio, T., Zhou, Y., Zhang, J., Zon, L., and Nimer, S.D. 2002. Isolation and characterization of runxa and runxb, zebrafish members of the runt family of transcriptional regulators. *Exp Hematol* **30**:1381–1389.

Burns, C.E., Traver, D., Mayhall, E., Shepard, J.L., and Zon, L.I. 2005. Hematopoietic stem cell fate is established by the Notch-Runx pathway. *Genes Dev* **19**:2331–2342.

Cantor, A.B. 2005. GATA transcription factors in hematologic disease. *Int J Hematol* **81**:378–384.

Davidson, A.J. and Zon, L.I. 2006. The caudal-related homeobox genes cdx1a and cdx4 act redundantly to regulate hox gene expression and the formation of putative hematopoietic stem cells during zebrafish embryogenesis. *Dev Biol* **292**:506–518.

Davidson, A.J., Ernst, P., Wang, Y., et al. 2003. cdx4 mutants fail to specify blood progenitors and can be rescued by multiple hox genes. *Nature* **425**:300–306.

Dieterlen-Lievre, F. 1974. [The origin of definitive hematopoetic stem cells in bird embryos: experimental analysis using quail-chicken chimeras.] *C R Acad Sci Hebd Seances Acad Sci D* **279**:915–918.

Dieterlen-Lievre, F. and Cumano, A. 1998. Cellular and molecular events that govern the development of the hematopoietic and immune system in the embryo. *Dev Comp Immunol* **22**:249–252.

Donovan, A., Brownlie, A., Zhou, Y., et al. 2000. Positional cloning of zebrafish ferroportin1 identifies a conserved vertebrate iron exporter. [See comment.] *Nature* **403**:776–781.

Evans, T. and Felsenfeld, G. 1989. The erythroid-specific transcription factor Eryf1: a new finger protein. *Cell* **58**:877–885.

Evans, T., Reitman, M., and Felsenfeld, G. 1988. An erythrocyte-specific DNA-binding factor recognizes a regulatory sequence common to all chicken globin genes. *Proc Natl Acad Sci U S A* **85**:5976–5980.

Galloway, J.L. and Zon, L.I. 2003. Ontogeny of hematopoiesis: examining the emergence of hematopoietic cells in the vertebrate embryo. *Curr Top Dev Biol* **53**:139–158.

Galson, D.L. and Housman, D.E. 1988. Detection of two tissue-specific DNA-binding proteins with affinity for sites in the mouse beta-globin intervening sequence 2. *Mol Cell Biol* **8**:381–392.

Glaser, S., Metcalf, D., Wu, L., et al. 2006. Enforced expression of the homeobox gene Mixl1 impairs hematopoietic differentiation and results in acute myeloid leukemia. *Proc Natl Acad Sci U S A* **103**:16460–16465.

Gumucio, D.L., Rood, K.L., Gray, T.A., Riordan, M.F., Sartor, C.I., and Collins, F.S. 1988. Nuclear proteins that bind the human gamma-globin gene promoter: alterations in binding produced by point mutations associated with hereditary persistence of fetal hemoglobin. *Mol Cell Biol* **8**:5310–5322.

Guo, W., Chan, A.P., Liang, H., et al. 2002. A human Mix-like homeobox gene MIXL shows functional similarity to Xenopus Mix.1. *Blood* **100**:89–95.

Hart, A.H., Hartley, L., Sourris, K., et al. 2002. Mixl1 is required for axial mesendoderm morphogenesis and patterning in the murine embryo. *Development* **129**:3597–3608.

Kau, C.L. and Turpen, J.B. 1983. Dual contribution of embryonic ventral blood island and dorsal lateral plate mesoderm during ontogeny of hemopoietic cells in Xenopus laevis. *J Immunol* **131**:2262–2266.

Knezetic, J.A. and Felsenfeld, G. 1989. Identification and characterization of a chicken alpha-globin enhancer. *Mol Cell Biol* **9**:893–901.

Liao, E.C., Trede, N.S., Ransom, D., Zapata, A., Kieran, M., and Zon, L.I. 2002. Non-cell autonomous requirement for the bloodless gene in primitive hematopoiesis of zebrafish. *Development* **129**:649–659.

Mantovani, R., Malgaretti, N., Giglioni, B., et al. 1987. A protein factor binding to an octamer motif in the gamma-globin promoter disappears upon induction of differentiation and hemoglobin synthesis in K562 cells. *Nucleic Acids Res* **15**:9349–9364.

Martin, D.I., Zon, L.I., Mutter, G., and Orkin, S.H. 1990. Expression of an erythroid transcription factor in megakaryocytic and mast cell lineages. *Nature* **344**:444–447.

Mead, P.E., Brivanlou, I.H., Kelley, C.M., and Zon, L.I. 1996. BMP-4-responsive regulation of dorsal-ventral patterning by the homeobox protein Mix.1. *Nature* **382**:357–360.

Mignotte, V., Wall, L., deBoer, E., Grosveld, F., and Romeo, P.H. 1989. Two tissue-specific factors bind the erythroid promoter of the human porphobilinogen deaminase gene. *Nucleic Acids Res* **17**:37–54.

Mohn, D., Chen, S.W., Dias, D.C., et al. 2003. Mouse Mix gene is activated early during differentiation of ES and F9 stem cells and induces endoderm in frog embryos. *Dev Dyn* **226**:446–459.

Mullins, M.C. and Nusslein-Volhard, C. 1993. Mutational approaches to studying embryonic pattern formation in the zebrafish. *Curr Opin Genet Dev* **3**:648–654.

Njajou, O.T., Vaessen, N., Joosse, M., et al. 2001. A mutation in SLC11A3 is associated with autosomal dominant hemochromatosis. *Nat Genet* **28**: 213–214.

North, T.E., Goessling, W., Walkley, C.R., et al. 2007. Prostaglandin E2 regulates vertebrate haematopoietic stem cell homeostasis. *Nature* **447**: 1007–1011.

Orkin, S.H. 1990. Globin gene regulation and switching: circa 1990. *Cell* **63**:665–672.

Pardanaud, L., Luton, D., Prigent, M., Bourcheix, L.M., Catala, M., and Dieterlen-Lievre, F. 1996. Two distinct endothelial lineages in ontogeny, one of them related to hemopoiesis. *Development* **122**:1363–1371.

Perkins, N.D., Nicolas, R.H., Plumb, M.A., and Goodwin, G.H. 1989. The purification of an erythroid protein which binds to enhancer and promoter elements of haemoglobin genes. *Nucleic Acids Res* **17**:1299–1314.

Plumb, M., Frampton, J., Wainwright, H., et al. 1989. GATAAG; a cis-control region binding an erythroid-specific nuclear factor with a role in globin and non-globin gene expression. *Nucleic Acids Res* **17**:73–92.

Ransom, D.G., Haffter, P., Odenthal, J., et al. 1996. Characterization of zebrafish mutants with defects in embryonic hematopoiesis. *Development* **123**: 311–319.

Robb, L., Hartley, L., Begley, C.G., et al. 2000. Cloning, expression analysis, and chromosomal localization of murine and human homologues of a Xenopus mix gene. *Dev Dyn* **219**:497–504.

Shaw, G.C., Cope, J.J., Li, L., et al. 2006. Mitoferrin is essential for erythroid iron assimilation. *Nature* **440**:96–100.

Smith, W.C. and Harland, R.M. 1991. Injected Xwnt-8 RNA acts early in Xenopus embryos to promote formation of a vegetal dorsalizing center. *Cell* **67**:753–765.

Stern, H.M., Murphey, R.D., Shepard, J.L., et al. 2005. Small molecules that delay S phase suppress a zebrafish bmyb mutant. *Nat Chem Biol* **1**:366–370.

Traver, D., Paw, B.H., Poss, K.D., Penberthy, W.T., Lin, S., and Zon, L.I. 2003. Transplantation and in vivo imaging of multilineage engraftment in zebrafish bloodless mutants. *Nat Immunol* **4**:1238–1246.

Traver, D., Winzeler, A., Stern, H.M., et al. 2004. Effects of lethal irradiation in zebrafish and rescue by hematopoietic cell transplantation. *Blood* **104**:1298–1305.

Turpen, J.B. and Smith, P.B. 1985. Dorsal lateral plate mesoderm influences proliferation and differentiation of hemopoietic stem cells derived from ventral lateral plate mesoderm during early development of Xenopus laevis embryos. *J Leukoc Biol* **38**:415–427.

Wall, L., deBoer, E., and Grosveld, F. 1988. The human beta-globin gene 3' enhancer contains multiple binding sites for an erythroid-specific protein. *Genes Dev* **2**:1089–1100.

Willey, S., Ayuso-Sacido, A., Zhang, H., et al. 2006. Acceleration of mesoderm development and expansion of hematopoietic progenitors in differentiating ES cells by the mouse Mix-like homeodomain transcription factor. *Blood* **107**:3122–3130.

Wingert, R.A., Galloway, J.L., Barut, B., et al. 2005. Deficiency of glutaredoxin 5 reveals Fe-S clusters are required for vertebrate haem synthesis. *Nature* **436**:1035–1039.

Wood, W.G. and Weatherall, D.J. 1983. Developmental genetics of the human haemoglobins. *Biochem J* **215**:1–10.

Zhao, Y., Ratnayake-Lecamwasam, M., Parker, S.K., et al. 1998. The major adult alpha-globin gene of antarctic teleosts and its remnants in the hemoglobin-less icefishes. Calibration of the mutational clock for nuclear genes. *J Biol Chem* **273**:14745–14752.

Zon, L.I., Mather, C., Burgess, S., Bolce, M.E., Harland, R.M., and Orkin, S.H. 1991. Expression of GATA-binding proteins during embryonic development in Xenopus laevis. *Proc Natl Acad Sci U S A* **88**:10642–10646.

TRACKING THE ROAD FROM INFLAMMATION TO CANCER: THE CRITICAL ROLE OF IκB KINASE (IKK)

MICHAEL KARIN

Laboratory of Gene Regulation and Signal Transduction, Departments of Pharmacology and Pathology, Cancer Center, School of Medicine, University of California, San Diego, La Jolla, California

I. Introduction

An association between inflammation and cancer has already been noted in the 19th century (Balkwill and Mantovani, 2001), and more recent epidemiological data had led to the estimate that at least 20% of cancer deaths are linked to chronic infections and persistent inflammation (Kuper et al., 2000). Notable examples are gastric cancer and *Helicobacter pylori* infections (Roder, 2002), hepatocellular carcinoma (HCC) and viral hepatitis (Fattovich et al., 2004), and colitis-associated cancer (CAC) (Ekbom, 1998). However, epidemiological associations do not establish causality, and the mechanisms that link inflammation and cancer have only been addressed recently. Initial work in this area had demonstrated the requirement of tumor necrosis factor (TNF)-α and signaling via the type I TNF-α receptor (TNFR1) in the development of squamous cell carcinoma (SCC) induced by the classical two-stage carcinogenesis protocol (Moore et al., 1999). The molecular mechanism by which TNFR1 signaling promotes tumor development, however, has not been thoroughly explored, although it is thought to be mediated via protein kinase C (PKC) α and activator protein 1 (AP-1) transcription factors (Arnott et al., 2002). Another cytokine, colony stimulating factor-1 (CSF-1), was shown to be required for progression of fully malignant mammary carcinoma in mice, presumably through its effect on macrophage development (Lin et al., 2001). The mechanisms by which macrophages stimulate the development and progression of mammary carcinoma are still being

The Harvey Lectures, Series 102, pages 133–152
©2010 by John Wiley & Sons, Inc.

unraveled (Pollard, 2004). Another inflammatory mediator, the enzyme cyclooxygenase 2 (COX2), responsible for inducible prostanoid synthesis (Garber, 2004), was shown to be required for the development of colonic polyps and adenomas in $Apc^{+/min}$ mice (Oshima et al., 1996). Furthermore, selective and non-selective COX2 inhibitors, including non-steroidal anti-inflammatory drugs (NSAIDs), have shown the ability, in large-scale clinical trials, to prevent the progression of adenomatous polyposis coli (APC) to colorectal adenocarcinoma, as well as reduce the incidence of colorectal cancer (Steinbach et al., 2000; Arnott et al., 2002; Koehne and Dubois, 2004). The mechanisms by which COX2 contributes to the progression of colorectal cancer appear to be mainly related to the synthesis of Prostaglandin E2 (PGE_2), which in turn stimulates angiogenesis and other processes (Koehne and Dubois, 2004).

Trying to explore and explain the molecular underpinnings that connect inflammation to cancer, we proposed in 2002 that transcription factor NF-κB, formed through the combinatorial dimerization of five family members in mammals (Ghosh and Karin, 2002), is the key molecular lynchpin that connects the two (Karin et al., 2002). This proposal was based on a large body of circumstantial evidence, documenting the presence of constitutively activated NF-κB in a large number of solid malignancies (Karin et al., 2002) and the well-established ability of NF-κB to up-regulate expression of key pro-inflammatory cytokines and enzymes, including TNF-α, IL-1, IL-6, CSF-1 and COX-2 (Barnes and Karin, 1997), as well as a number of genes that code for anti-apoptotic proteins (Liu et al., 1996; Lin and Karin, 2003). We thus proposed that persistent infections and chronic inflammation lead to NF-κB activation, and once induced, NF-κB target genes protect pre-neoplastic and fully malignant cells from apoptosis induced by genomic surveillance mechanisms (e.g. p53) or genotoxic anti-cancer drugs and radiation. Thus, NF-κB activation is likely to not only contribute to the emergence and expansion of pre-neoplastic cells but can also confer drug and radiation resistance upon fully developed tumors. In the latter context, it is worth mentioning that we recently found that in addition to inhibition of apoptosis, NF-κB can also prevent necrotic cell death (Kamata et al., 2005), a form of cell death that may be more relevant to the mode of action of many anti-cancer drugs (Zong and Thompson, 2006).

The mechanisms responsible for NF-κB activation in solid malignancies (mainly carcinomas) are still not fully clear, but in most cases are unlikely to be due to mutations intrinsic to the cancer cell itself (Karin et al., 2002). One possible way in which NF-κB becomes activated is in response to production of pathogen-associated molecular patterns (PAMPs) in the case of persistent microbial and viral infections or through pro-inflammatory cytokines such as TNF-α and IL-1, during chronic inflammation. All of these agents act via diverse cell surface receptors that lead to the activation of a common molecular target – the IκB kinase (IKK) complex (DiDonato et al., 1997; Rothwarf and Karin, 1999). The IKK complex consists of two catalytic subunits: IKKα and IKKβ, and a regulatory subunit IKKγ/NEMO. Gene targeting experiments revealed that activation of NF-κB in response to PAMPs and pro-inflammatory cytokines is absolutely dependent on IKKγ (Makris et al., 2000). Similar experiments indicate that although IKKα and IKKβ are 52% identical in sequence, IKKβ plays a much more critical role in the response to PAMPs and pro-inflammatory cytokines than IKKα (Hu et al., 1999; Li et al., 1999; Chen et al., 2003). By contrast, IKKα, but not IKKβ, kinase activity is required for activation of an alternative NF-κB signaling pathway based on processing of NF-κB2/p100:RelB complexes to NF-κB2/p52:RelB dimers (Bonizzi and Karin, 2004). In addition, IKKα, and not IKKβ, is required for formation of stratified epithelia, such as the epidermis, but this function does not depend on its protein kinase activity (Hu et al., 1999, 2001; Sil et al., 2004). Given the critical role of IKKβ in the activation of the classical NF-κB signaling pathway in response to PAMPs and pro-inflammatory cytokines, we used targeted disruptions of the *Ikkβ* gene to study the role of IKKβ-dependent NF-κB activation in a variety of cancer models in mice.

II. THE SIMPLE: THE ROLE OF EPITHELIAL AND MYELOID NF-κB IN CAC

We first examined the cancer-promoting role of IKKβ in a mouse model of CAC, based on the application of azoxymethane (AOM), a procarcinogen that undergoes metabolic activation in colonic epithelial cells, and dextran sulfate sodium salt (DSS), an irritant that causes colonic inflammation (Okayasu et al., 1996). Application of either AOM or DSS alone is not sufficient for effective tumor induction, but a combination of both agents results in efficient formation of adenomas and adenocarcinomas with 100% penetrance in C57BL6 mice. As in human CAC, these tumors develop at the distal part of the colon. Using mice homozygous for a "floxed" $Ikk\beta^F$ allele (Chen et al., 2003) that have been genetically crossed to *Villin-Cre* mice, in which Cre recombinase is specifically expressed in intestinal epithelial cells (IEC), we first examined the contribution of IKKβ in IEC to CAC development. Importantly, $Ikk\beta^{F/F}$/*Villin-Cre* mice ($Ikk\beta^{\Delta IEC}$) are healthy and exhibit normal colon structure and function unless challenged (Chen et al., 2003). Strikingly, $Ikk\beta^{\Delta IEC}$ mice were found to exhibit an impressive 80% decline in CAC load relative to similarly treated wild-type (WT) mice, thereby providing the first conclusive evidence for the role of IKKβ-dependent NF-κB activation in the development of an inflammation-promoted cancer (Greten et al., 2004). $Ikk\beta^{\Delta IEC}$ mice lack IKKβ in the very same cells that are subject to the mutagenic activity of AOM and give rise to the genetically transformed component of CAC – the malignant adenocarcinoma cell. However, another important cell type in inflammation-promoted cancer is the macrophage, a critical source of inflammatory cytokines and mediators (Coussens and Werb, 2002; Balkwill et al., 2005). To investigate the role of IKKβ in macrophages in CAC development, we introduced into the $Ikk\beta^{F/F}$ strain a Cre transgene driven by the LysM promoter, which is active in mature macrophages and neutrophils (Clausen et al., 1999). $Ikk\beta^{F/F}$/*LysM-Cre* mice ($Ikk\beta^{\Delta mye}$), which are also healthy and physiologically normal when kept unchallenged, were found to exhibit a 50% decrease in tumor multiplicity, but a greater decline in total tumor load, as most of the tumors present in these mice were smaller in size relative to tumors in WT mice (Greten et al., 2004). These results provided the first clear-cut, genetically supported evidence for the tumor-promoting role of NF-κB activation in inflammatory cells, in this case, the lamina propria macrophage or a dendritic cell. Although both $Ikk\beta^{\Delta IEC}$ and $Ikk\beta^{\Delta mye}$ mice display reduced tumor development, detailed analysis revealed that in each cell type, IKKβ-driven NF-κB accomplishes its tumor-promoting function through a different mechanism. In the IEC, the most important tumor-promoting function of NF-κB is to endow the emerging premalignant cell with a survival advantage by inducing expression of anti-apoptotic genes, such as $Bcl-X_L$, whose products prevent the elimination of the transformed cell through genomic surveillance mechanisms. It should be noted, however, that the IKKβ deficiency in the IEC had no effect on cell proliferation or the spectrum of oncogenic mutations induced by AOM, many of which occur in the β catenin (*Catnb*) gene (Greten et al., 2004). In myeloid cells, however, IKKβ-driven NF-κB promotes tumor develoment through the induction of growth factors that stimulate the proliferation of pre-neoplastic cells (Greten et al., 2004). One of these factors was suggested to be IL-6 (Becker et al., 2004). We recently confirmed the important role of IL-6, which is encoded by a typical NF-κB target gene, in CAC development, and reaffirmed our earlier findings that during the initial stages of the CAC protocol, the main site of IL-6 production is lamina propria myeloid cell (Grivennikov et al., 2009).

In summary, the studies on IKKβ in CAC have provided critical support for the important tumor-promoting function of NF-κB. These studies have also demonstrated for the first time that IKKβ-driven NF-κB mostly affects tumor promotion rather than tumor initiation through distinct mechanisms in different cell types. In this case, NF-κB activation in IEC prevents the elimination of pre-neoplastic cells by genomic surveillance mechanisms, whereas NF-κB in lamina propria inflammatory cells induces the production of growth factors, such as IL-6, that stimulate the proliferation of premalignant cells (Fig. 7.1).

III. The Complex: The Unexpected Role of Hepatocyte IKKβ in Chemically Induced HCC

One of the most common types of inflammation-linked cancers worldwide is HCC (Fattovich et al., 2004). Unfortunately, however, the viruses that greatly enhance the risk of HCC, Hepatitis B and C Virus (HBV and HCV) cannot be easily propagated in mice and therefore cannot be used to induce HCC in this genetically manipulatable small mammal. Although liver-targeted expression of various oncogenes can lead to HCC development in mice (Fausto, 1999; Lewis et al., 2005), we chose a different model based on the treatment with a chemical carcinogen, diethyl nitrosamine (DEN), that induces HCC whose gene expression profile is very similar to that of aggressive human HCC (Thorgeirsson and Grisham, 2002). Despite the similar gene expression pattern, we did not anticipate that the carcinogenic action of DEN, which forms a very potent alkylating agent upon metabolic activation in hepatocytes (Sarma et al., 1986), depends on inflammation or inflammatory processes, which have an important role in HCC induced by chronic viral hepatitis (Chisari, 1995; Bosch et al., 2004). Based on the ability of DEN to induce DNA damage

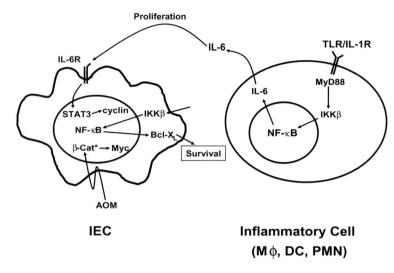

Fig. 7.1.   IKKβ-driven NF-κB promotes the development of CAC by acting in two different cell types. NF-κB activation in premalignant intestinal epithelial cells (IEC carrying oncogenic mutations induced by AOM exposure) results in up-regulation of pro-survival genes such as Bcl-X$_L$, thereby preventing apoptotic elimination via genomic surveillance mechanisms. IKKβ-driven NF-κB also contributes to CAC development in inflammatory cells (macrophages (Mφ), dendritic cells (DC), and/or neutrophils (PMN) of the lamina propria) where it induces expression of epithelial cell growth factors, such as IL-6.

and thereby activate the p53-dependent cytotoxic stress response and the previously documented ability of NF-κB to counteract the pro-apoptotic function of p53 (Tergaonkar et al., 2002), we expected that mice lacking IKKβ in hepatocytes, so-called $Ikk\beta^{\Delta hep}$ mice (Maeda et al., 2003), will exhibit a more efficient p53-promoted apoptotic response to DEN and therefore should exhibit reduced HCC load relative to DEN-treated WT mice. Counter to these expectations, $Ikk\beta^{\Delta hep}$ mice were found to develop many more and faster-growing tumors than control mice (Maeda et al., 2005). However, HCC, induced in $Ikk\beta^{+/-}$ mice, did not display the loss of heterozygocity, thereby indicating that although IKKβ in hepatocytes is an inhibitor of HCC development, it does not act as a classical tumor suppressor. Despite the surprising increase in HCC development, $Ikk\beta^{\Delta hep}$

mice did exhibit the expected increase in DEN-induced apoptosis as well as necrosis, and, due to its highly efficient regenerative capacity, the liver of $Ikk\beta^{\Delta hep}$ mice contained many more proliferating cells several days after DEN exposure than the liver of similarly treated WT (or $Ikk\beta^{F/F}$) mice (Maeda et al., 2005). Dual-labeling experiments revealed that most of these proliferating cells were located next to dying cells, a classic example of compensatory proliferation. However, further analysis revealed that neither IKKβ nor NF-κB is a negative regulator of the hepatocyte cell cycle and that the increased compensatory proliferation seen in $Ikk\beta^{\Delta hep}$ mice is directly due to the elevated level of hepatocyte death in these animals. NF-κB activation promotes hepatocyte survival through a variety of mechanisms, including the induction of anti-apoptotic proteins, such as Bcl-X_L. However, DEN induces both apoptosis and necrosis through a complex cytotoxic mechanism that, in addition to the induction of DNA damage, includes the generation of reactive oxygen species (ROS). We found that NF-κB activation can attenuate ROS accumulation through the induction of antioxidants such as Mn superoxide dismutase or SOD2 (Kamata et al., 2005), whereas others have shown a role for NF-κB-induced ferritin heavy chain (FHC) in ROS elimination (Pham et al., 2004). Indeed, $Ikk\beta^{\Delta hep}$ mice display elevated accumulation of ROS in their hepatocytes after DEN administration relative to similarly treated WT mice (Maeda et al., 2005). Elevated ROS accumulation contributes to DEN-induced cell death and tumor development, as both are inhibited in response to the administration of butylated hydroxyanisole (BHA), a potent antioxidant. Although it is possible that elevated ROS accumulation may lead to higher levels of DNA damage and oncogenic mutations, the main effect of ROS is likely to be enhanced cell death. One of the ways in which ROS accumulation enhances cell death is through the oxidation of a critical cysteine residue in the catalytic pocket of MAPK (MAP Kinase) phosphatases (MKPs) (Kamata et al., 2005). This results in the inhibition of MKP activity and the sustained activation of different MAPKs, including JNK1, a critical contributor to DEN-induced hepatocyte death (Kamata et al., 2005; Sakurai et al., 2006). Importantly, the inactivation of JNK1 attenuated DEN-induced hepatocyte death and greatly reduced the occurrence of DEN-induced HCC (Sakurai et al., 2006).

Collectively, these results indicate that the major role of IKKβ-driven NF-κB in hepatocytes is to provide protection against a variety of

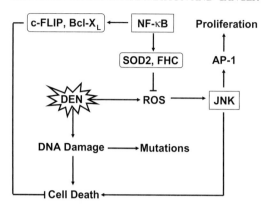

Fig. 7.2.   IKKβ and JNK1 inversely control hepatocyte survival and compensatory proliferation in DEN-treated mice. DEN undergoes metabolic activation in zone 3 hepatocytes resulting in accumulation of ROS, which exert a cytotoxic effect that may be due to sustained JNK activation. DEN can also lead to necrotic cell death through induction of DNA damage and also causes IKK and JNK (Jun Kinase) activation through unknown mechanisms. Activation of NF-κB promotes cell survival through different mechanisms, including the up-regulation of antioxidants, such as SOD2 and FHC that prevent excessive ROS accumulation and prolonged JNK activation and induction of anti-apoptotic proteins, such as c-FLIP and Bcl-X_L. Insufficient activation of NF-κB promotes ROS accumulation, leading to sustained JNK activation and cell death. In addition to its role in cell death, JNK can activate AP-1 transcription factors and enhance cyclin D expression, thereby promoting the proliferation of surviving hepatocytes.

cytotoxic challenges, and thereby inhibit injury-induced inflammation and compensatory proliferation (Fig. 7.2). These results provide strong biochemical and molecular genetic support to the notion that HCC, whether induced by the administration of a carcinogen (DEN) in mice or by chronic viral hepatitis in humans, is the end result of repetitive liver injury followed by regenerative cell proliferation. High rates of compensatory proliferation increase the likelihood that oncogenic mutations and DNA rearrangements will be genetically fixed, transmitted to subsequent generations and eventually lead to tumor development. This general mechanism may be applicable to other tissues with inherently low rates of cell division but high capacity for compensatory proliferation in response to injury.

IV. The Sexy: Role of Inflammation and Gender in Hepatocarcinogenesis

Although the inactivation of IKKβ in hepatocytes enhanced the development of HCC by setting a lower threshold for the induction of liver injury, the inactivation of IKKβ in inflammatory cells inhibited the development of HCC, just as it did for CAC (Maeda et al., 2005). These results indicated that even though DEN-induced hepatocarcinogenesis was not known to involve inflammation, it is dependent on inflammatory signaling, in this case, NF-κB activation in hematopoietic-derived cells, after all. We speculated that the most critical inflammatory cell for HCC development is the Kupffer cell (KC), the resident liver macrophage (Maeda et al., 2005). In support of this proposal, the KC is a major source of IL-6 production following DEN administration (Maeda et al., 2005), and IL-6 is essential for DEN-induced hepatocarcinogenesis (Naugler et al., 2007). We have also proposed that necrotic injury to hepatocytes results in the release of inflammatory mediators that lead to KC activation (Maeda et al., 2005). As the identity of the hypothetical mediators released by dying hepatocytes and the signaling mechanisms through which they operate were nebulous, this proposal remained purely speculative until recently.

While trying to find ways and means to examine and understand the putative connection between dying hepatocytes and KC in the context of DEN-induced HCC, we addressed a different and seemingly unrelated problem – the gender bias in hepatocarcinogenesis. The worldwide incidence if HCC is three to five times higher among males than among females (Bosch et al., 2004). This gender bias becomes even more staggering when the incidence of HCC is compared in individuals who are younger than 50 years of age, a group that includes pre-menopausal women. In that group, the incidence of HCC is 7–10 times higher in men than in women. A similar gender bias is seen in rodent models of HCC (Nakatani et al., 2001). We noted, for instance, that following a single administration of DEN at 2 weeks of age, the incidence of HCC in 8-month-old male mice (100%) is at least six times higher than in females of the same age (15%). This difference declined to 2.5-fold in $Ikk\beta^{\Delta hep}$ female mice (Maeda et al., 2005). These findings suggested that the refractoriness of female mice to DEN-induced HCC may be related to their resistance to DEN-induced hepatic injury, which is reduced upon the deletion of hepatocyte IKKβ. Indeed, we found that male mice develop much more extensive liver injury than females after administration of either DEN or a different liver toxin – carbon tetrachloride (CCl$_4$), a liver tumor promoter (Naugler et al., 2007). Although CCl$_4$ is not a mutagenic carcinogen, it undergoes metabolic activation and leads to ROS accumulation in the same type of cell as DEN – the zone 3 hepatocyte. Importantly, the administration of either DEN or CCl$_4$ results in a much higher IL-6 production in male mice than in females, and the ablation of IL-6 reduces hepatic injury and almost completely prevents the development of HCC in male mice (Naugler et al., 2007). Thus, in the absence of IL-6, male mice do not develop any more HCCs than female mice. As mentioned above, IL-6 is mainly produced by KC, and the incubation of KC with proteins released by necrotic hepatocytes results in IL-6 production that depends on IKKβ activation in KC and is suppressible by estrogen (E$_2$) and pure estrogen receptor α (ERα) agonists.

We speculated that proteins released by necrotic hepatocytes may activate the KC via innate immune receptors of the Toll/IL-1R family and therefore examined the dependence of this response on MyD88, an adaptor protein that plays a critical role in the IKK activation by this family of receptors (Akira et al., 2006). Indeed, KC isolated from $Myd88^{-/-}$ mice produced very little IL-6 upon incubation with the products of hepatocyte necrosis. Most importantly, the administration of either DEN or CCl$_4$ to $Myd88^{-/-}$ male mice resulted in very little IL-6 production and $Myd88^{-/-}$ males exhibited a fivefold decrease in HCC multiplicity relative to similarly treated WT males (Naugler et al., 2007).

These findings suggest the following mechanism for induction of HCC by DEN and probably by other hepatic carcinogens. DEN undergoes metabolic activation in zone 3 hepatocytes, a few of which may experience oncogenic mutations but many of which die. The dead hepatocytes release normal cellular constituents that lead to the activation of KC via TLR/IL-1R family members in a MyD88-dependent manner (Fig. 7.3). This results in the production of IL-6, which amplifies the injury response through yet-to-be identified mechanisms and stimulates compensatory hepatocyte proliferation probably through the activation of the STAT3 transcription factor (Naugler et al., 2007). Thus, DEN-induced hepatocarcinogenesis depends on an inflammatory interplay between dying hepatocytes, KC and living hepatocytes that carry oncogenic mutations

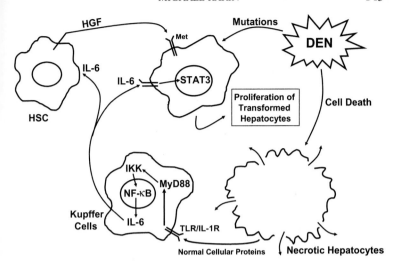

Fig. 7.3. An inflammatory link between hepatocyte necrosis and Kupffer cell activation controls compensatory proliferation and development of hepatocellular carcinoma (HCC). Proteins released by necrotic hepatocytes activate Kupffer cells via members of the Toll/IL-1R family leading to induction of cytokines such as IL-6. IL-6 activates STAT3 in surviving hepatocytes and leads to activation of hepatic stellate cells (HSC) that produce hepatocyte growth factor (HGF). Collectively, these cytokines stimulate the proliferation of surviving hepatocytes. Proliferating hepatocytes that carry oncogenic mutations induced by the mutagenic activity of DEN proceed to form HCC.

(Fig. 7.3). The same general mechanism may account for the development of human HCC, and therefore our results suggest that ERα antagonists that are capable of inhibiting IL-6 production without inducing feminization may be used as chemopreventive agents that block the progression of hepatitis to HCC. Alternatively, a similar effect may be produced by drugs that interfere with IL-6 signaling. Most recently, we found that the major inflammatory mediator released by necrotic hepatocytes is IL-Iα (Sukurai et al., 2008).

V. The Different: IKKα as an Enhancer of Prostate Cancer Metastasis

The findings described above in both CAC and chemically induced HCC not only underscore the important tumor-promoting function of

IKKβ in inflammatory cells but also reveal the complex effects of IKKβ-dependent NF-κB activation in epithelial cells on tumor development. These studies also revealed a critical role for inflammation in early tumor promotion (Karin and Greten, 2005). However, the effect of inflammation is probably not limited to tumor promotion, and it may also affect neoplastic progression and the formation of distant site metastases. As neither the CAC nor the HCC models described above are suitable for studying metastasis, we have turned to the TRAMP model for prostate carcinoma (CaP), which mimics many of the aspects of malignant and metastatic progression seen in human CaP, including the delayed but frequent development of distal organ metastases (Greenberg et al., 1995; Kaplan-Lefko et al., 2003). First, we examined the role of IKKβ in prostate epithelial cells on the formation of CaP in TRAMP mice, in which tumors are induced through prostate-specific expression of SV40 T antigen, but no effect on either metastatogenesis or tumorigenesis was found (Ammirante et al., 2009). We, therefore, have switched to examine the role of IKKα in the development and progression of this malignancy. To that end, we have used the $Ikk\alpha^{AA}$ knock-in mouse in which the two serine phosphorylation sites in the activation loop of IKKα were replaced with alanines (Cao et al., 2001). These mice express normal amounts of IKKα whose kinase activity cannot be turned on in response to upstream stimuli. Thus, kinase-independent functions of IKKα, for instance in skin development (Hu et al., 2001), remain intact in $Ikk\alpha^{AA}$ mice, while those that depend on kinase activation are absent. We found that TRAMP mice rendered homozygous for the $Ikk\alpha^{AA}$ mutation exhibited decelerated tumor development, but eventually, all died of primary CaP. However, when analyzed at the time of death, the mutant mice exhibited much fewer secondary site metastases than WT/TRAMP mice (Luo et al., 2007). Subsequent analysis verified that IKKα kinase activity was required for metastatogenesis but was not required for the formation of primary tumors and had no effect on their tumorigenic potential, measured by subcutaneous implantation. To understand how IKKα controls metastatic activity, we compared the expression of 40 known positive and negative regulators of metastasis at different stages of tumor progression between WT/TRAMP and $Ikk\alpha^{AA}$/TRAMP mice and found that IKKα kinase activity controlled the expression of only a single metastasis regulating gene coding for maspin (Luo et al., 2007). Maspin is a serine protease inhibitor originally identified by its anti-metastatic activity in mammary carcinomas (Zou et al., 1994). At the early stages of CaP development, *WT*/TRAMP and *Ikkα*[AA]/TRAMP express similar amounts of maspin in adenocarcinoma cells, but at later stages, WT/*TRAMP* mice exhibit loss of maspin in CaP cells, and this decline correlates with the appearance of secondary site metastases. *Ikkα*[AA]/TRAMP mice, however, retain high levels of maspin expression in CaP cells throughout tumor progression and correspondingly exhibit very few metastases (Luo et al., 2007). IKKα controls maspin expression at the transcriptional level, and this requires its entry into the nucleus. Curiously, IKKα contains a nuclear localization sequence that is not present in IKKβ (Sil et al., 2004). Whereas normal prostate epithelial cells or early CaP cells contain little, if any, IKKα in the nucleus, advanced WT CaP cells isolated from mice that exhibit metastases contain significant amounts of nuclear IKKα (Luo et al., 2007). In fact, the amount of nuclear IKKα is directly related to metastatic progression and inversely related to maspin expression not only in TRAMP mice but also in human CaP patients (Luo et al., 2007). Exactly how nuclear IKKα controls maspin transcription is not fully clear, but chromatin immunoprecipitation experiments indicate that IKKα is recruited to the maspin promoter.

We examined which signals trigger nuclear translocation of IKKα and found a good correlation between reduced maspin expression, nuclear accumulation of IKKα and the presence of infiltrating T cells and macrophages, which appear only at late stages of tumor progression (Luo et al., 2007). We also found that advanced prostate tumors contained 50-fold more receptor activator of NF-κB (RANK) ligand (RANKL) mRNA and 20-fold more lymphotoxin α (LTα) mRNA than early tumors that contain little, if any, nuclear IKKα. In vitro, RANKL, which activates NF-κB through RANK, a member of the TNF receptor family, induced nuclear translocation of IKKα in WT prostate epithelial cells and down-regulated maspin expression. Nonetheless, this effect of IKKα was not mediated through NF-κB, and IKKβ, which does not translocate to the nucleus, was found to have no effect on maspin expression (Luo et al., 2007). Furthermore, RANKL had no effect on maspin expression in *Ikkα*[AA] prostate epithelial cells, indicating that down-regulation of maspin expression is IKKα dependent.

In summary, these results outline a new pathway through which tumor-induced inflammation, whose hallmark is the appearance of infiltrating inflammatory cells within the growing tumor, can stimulate metastatic

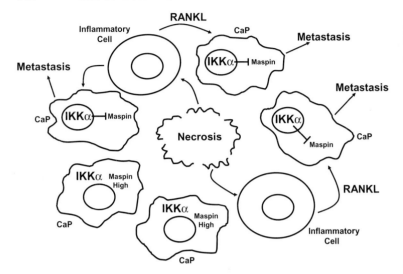

Fig. 7.4. A pathway based on IKKα activation and nuclear translocation controls metastatic progression in prostate cancer. The rapid growth of early prostate adenocarcinomas results in necrotic death of CaP cells located at the tumor's core that are starved for nutrients and oxygen. Necrotic cells release proteins that lead to recruitment and activation of inflammatory cells. The latter produce cytokines, such as RANKL, that activate IKKα and induce its nuclear translocation. Once a sufficient amount of activated IKKα had accumulated in the nucleus, maspin expression is repressed and, eventually, is epigenetically silenced. Absence of maspin expression results in metastasis.

progression in CaP and possibly also in mammary/breast cancer. We propose that accelerated tumor growth results in the necrotic death of a subpopulation of carcinoma cells that are being starved for nutrients and/or oxygen. Necrosis results in the recruitment and activation of inflammatory cells, macrophages and T cells that produce cytokines that further stimulate tumor growth and angiogenesis. Cytokines, such as RANKL, lead to the activation and nuclear translocation of IKKα, and this results in the repression of maspin expression. Eventually, the repressed maspin gene is permanently shutoff through epigenetic silencing. Cells that no longer express maspin gain full metastatic potential. This model, depicted in Figure 7.4, provides a good explanation for the long delay that is commonly associated with metastatic progression. It should be noted, however, that inflammation is likely to stimulate metastatogenesis through additional mechanisms that remain to be unraveled.

VI. Conclusions

Starting with the hypothesis that transcription factor NF-κB and the signaling pathways that stimulate its activity provide a molecular link between inflammation and cancer (Karin et al., 2002), my coworkers were able to provide ample support for the tumor-promoting function of IKKβ and its target NF-κB in several distinct models of cancer (Greten et al., 2004; Luo et al., 2004; Maeda et al., 2005; Sakurai et al., 2006; Naugler et al., 2007). We have also succeeded in identifying a new pathway in which IKKα plays a critical, but NF-κB-independent, role in the control of metastatogenesis. While some of the mechanistic details of the different pathways linking inflammation and cancer remain to be worked out, the major challenge in the near future is to validate these findings made in mice in human cancer patients and translate them to develop new and improved therapeutic and preventive strategies to reduce the burden of both primary and metastatic cancers. One advantage of targeting the inflammatory cell component of tumors is that such cells remain genetically normal and stable and thus are unlikely to develop drug resistance as easily as the genetically unstable carcinoma cells. However, we envision anti-inflammatory therapy as being most effective in combination with conventional cytotoxic cancer therapy.

References

Akira, S., Uematsu, S., and Takeuchi, O. 2006. Pathogen recognition and innate immunity. *Cell* **124**:783–801.

Ammirante, M., Luo, J.-L., Grivennikov, S., Nedospasov, S., Karin, M. (2009) B cell-derived lymphotoxin promotes castration – resistant prostate cancer. *Nature* in press.

Arnott, C.H., Scott, K.A., Moore, R.J., Hewer, A., Phillips, D.H., Parker, P., Balkwill, F.R., and Owens, D.M. 2002. Tumour necrosis factor-alpha mediates tumour promotion via a PKC alpha- and AP-1-dependent pathway. *Oncogene* **21**:4728–4738.

Balkwill, F. and Mantovani, A. 2001. Inflammation and cancer: back to Virchow? *Lancet* **357**:539–545.

Balkwill, F., Charles, K.A., and Mantovani, A. 2005. Smoldering and polarized inflammation in the initiation and promotion of malignant disease. *Cancer Cell* 7:211–217.

Barnes, P.J. and Karin, M. 1997. NF-kB – a pivotal transcription factor in chronic inflammatory diseases. *N Engl J Med* 336:1066–1071.

Becker, C., Fantini, M.C., Schramm, C., Lehr, H.A., Wirtz, S., Nikolaev, A., Burg, J., Strand, S., Kiesslich, R., Huber, S., Ito, H., Nishimoto, N., Yoshizaki, K., Kishimoto, T., Galie, P.R., Blessing, M., Rose-John, S., Neurath, M.F. 2004. TGF-beta suppresses tumor progression in colon cancer by inhibition of IL-6 trans-signaling. *Immunity* 21:491–501.

Bonizzi, G. and Karin, M. 2004. The two NF-kB activation pathways and their role in innate and adaptive immunity. *Trends Immunol* 25:280–288.

Bosch, F.X., Ribes, J., Diaz, M., and Cleries, R. 2004. Primary liver cancer: worldwide incidence and trends. *Gastroenterology* 127:S5–S16.

Cao, Y., Bonizzi, G., Seagroves, T.N., Greten, F.R., Johnson, R., Schmidt, E.V., and Karin, M. 2001. IKKa provides an essential link between RANK signaling and cyclin D1 expression during mammary gland development. *Cell* 107:763–775.

Chen, L.W., Egan, L., Li, Z.W., Greten, F.R., Kagnoff, M.F., and Karin, M. 2003. The two faces of IKK and NF-kB inhibition: prevention of systemic inflammation but increased local injury following intestinal ischemia-reperfusion. *Nat Med* 9:575–581.

Chisari, F.V. 1995. Hepatitis B virus transgenic mice: insights into the virus and the disease. *Hepatology* 22:1316–1325.

Clausen, B.E., Burkhardt, C., Reith, W., Renkawitz, R., and Forster, I. 1999. Conditional gene targeting in macrophages and granulocytes using LysMcre mice. *Transgenic Res* 8:265–277.

Coussens, L.M. and Werb, Z. 2002. Inflammation and cancer. *Nature* 420:860–867.

DiDonato, J.A., Hayakawa, M., Rothwarf, D.M., Zandi, E., and Karin, M. 1997. A cytokine-responsive IkB kinase that activates the transcription factor NF-kB. *Nature* 388:548–554.

Ekbom, A. 1998. Risk of cancer in ulcerative colitis. *J Gastrointest Surg* 2:312–313.

Fattovich, G., Stroffolini, T., Zagni, I., and Donato, F. 2004. Hepatocellular carcinoma in cirrhosis: incidence and risk factors. *Gastroenterology* 127:S35–S50.

Fausto, N. 1999. Mouse liver tumorigenesis: models, mechanisms, and relevance to human disease. *Semin Liver Dis* 19:243–252.

Garber, K. 2004. Aspirin for cancer chemoprevention: still a headache? *J Natl Cancer Inst* 96:252–253.

Ghosh, S. and Karin, M. 2002. Missing pieces in the NF-kB puzzle. *Cell* **109**(Suppl.):S81–S96.

Greenberg, N.M., DeMayo, F., Finegold, M.J., Medina, D., Tilley, W.D., Aspinall, J.O., Cunha, G.R., Donjacour, A.A., Matusik, R.J., and Rosen, J.M. 1995. Prostate cancer in a transgenic mouse. *Proc Natl Acad Sci U S A* **92**:3439–3443.

Greten, F.R., Eckmann, L., Greten, T.F., Park, J.M., Li, Z.W., Egan, L.J., Kagnoff, M.F., and Karin, M. 2004. IKKb links inflammation and tumorigenesis in a mouse model of colitis-associated cancer. *Cell* **118**:285–296.

Grivennikov, S., Karin, E., Terzic, J., Mucida, D., Yu, G.Y., Vallabhapurapu, S., Scheller, J., Rose-John, S., Cheroutre, H., Eckmann, L., Karin, M. (2009) IL-6 and Stat 3 are required for survival of intestinal epithelial cells and development of colitis-associated cancer. *Cancer Cell* **15**:103–113.

Hu, Y., Baud, V., Delhase, M., Zhang, P., Deerinck, T., Ellisman, M., Johnson, R., and Karin, M. 1999. Abnormal morphogenesis but intact IKK activation in mice lacking the IKKa subunit of the IkB kinase. *Science* **284**:316–320.

Hu, Y., Baud, V., Oga, T., Kim, K.I., Yoshida, K., and Karin, M. 2001. IKKa controls formation of the epidermis independently of NF-kB. *Nature* **410**:710–714.

Kamata, H., Honda, S., Maeda, S., Chang, L., Hirata, H., and Karin, M. 2005. Reactive oxygen species promote TNFa-induced death and sustained JNK activation by inhibiting MAP kinase phosphatases. *Cell* **120**:649–661.

Kaplan-Lefko, P.J., Chen, T.M., Ittmann, M.M., Barrios, R.J., Ayala, G.E., Huss, W.J., Maddison, L.A., Foster, B.A., and Greenberg, N.M. 2003. Pathobiology of autochthonous prostate cancer in a pre-clinical transgenic mouse model. *Prostate* **55**:219–237.

Karin, M. and Greten, F.R. 2005. NF-kB: linking inflammation and immunity to cancer development and progression. *Nat Rev Immunol* **5**:749–759.

Karin, M., Cao, Y., Greten, F.R., and Li, Z.W. 2002. NF-kB in cancer: from innocent bystander to major culprit. *Nat Rev Cancer* **2**:301–310.

Koehne, C.H. and Dubois, R.N. 2004. COX-2 inhibition and colorectal cancer. *Semin Oncol* **31**:12–21.

Kuper, H., Adami, H.O., and Trichopoulos, D. 2000. Infections as a major preventable cause of human cancer. *J Intern Med* **248**:171–183.

Lewis, B.C., Klimstra, D.S., Socci, N.D., Xu, S., Koutcher, J.A., and Varmus, H.E. 2005. The absence of p53 promotes metastasis in a novel somatic mouse model for hepatocellular carcinoma. *Mol Cell Biol* **25**:1228–1237.

Li, Z.-W., Chu, W., Hu, Y., Delhase, M., Deerinck, T., Ellisman, M., Johnson, R., and Karin, M. 1999. The IKKb subunit of IkB kinase (IKK) is essential for NF-kB activation and prevention of apoptosis. *J Exp Med* **189**:1839–1845.

Lin, A. and Karin, M. 2003. NF-kB in cancer: a marked target. *Semin Cancer Biol* **13**:107–114.

Lin, E.Y., Nguyen, A.V., Russell, R.G., and Pollard, J.W. 2001. Colony-stimulating factor 1 promotes progression of mammary tumors to malignancy. *J Exp Med* **193**:727–740.

Liu, Z.-G., Hu, H., Goeddel, D.V., and Karin, M. 1996. Dissection of TNF receptor 1 effector functions: JNK activation is not linked to apoptosis, while NF-kB activation prevents cell death. *Cell* **87**:565–576.

Luo, J.L., Maeda, S., Hsu, L.C., Yagita, H., and Karin, M. 2004. Inhibition of NF-kB in cancer cells converts inflammation-induced tumor growth mediated by TNFa to TRAIL-mediated tumor regression. *Cancer Cell* **6**:297–305.

Luo, J.L., Tan, W., Ricono, J.M., Korchynskyi, O., Zhang, M., Gonias, S.L., Cheresh, D.A., and Karin, M. 2007. Nuclear cytokine activated IKKa controls prostate cancer metastasis by repressing maspin. *Nature* **446**:690–694.

Maeda, S., Chang, L., Li, Z.W., Luo, J.L., Leffert, H., and Karin, M. 2003. IKKb is required for prevention of apoptosis mediated by cell-bound but not by circulating TNFalpha. *Immunity* **19**:725–737.

Maeda, S., Kamata, H., Luo, J.L., Leffert, H., and Karin, M. 2005. IKKb couples hepatocyte death to cytokine-driven compensatory proliferation that promotes chemical hepatocarcinogenesis. *Cell* **121**:977–990.

Makris, C., Godfrey, V.L., Krahn-Senftleben, G., Takahashi, T., Roberts, J.L., Schwarz, T., Feng, L., Johnson, R.S., and Karin, M. 2000. Female mice heterozygous for IKK gamma/NEMO deficiencies develop a dermatopathy similar to the human X-linked disorder incontinentia pigmenti. *Mol Cell* **5**:969–979.

Moore, R.J., Owens, D.M., Stamp, G., Arnott, C., Burke, F., East, N., Holdsworth, H., Turner, L., Rollins, B., Pasparakis, M., Kolias, G., Balkwill, F. 1999. Mice deficient in tumor necrosis factor-alpha are resistant to skin carcinogenesis. *Nat Med* **5**:828–831.

Nakatani, T., Roy, G., Fujimoto, N., Asahara, T., and Ito, A. 2001. Sex hormone dependency of diethylnitrosamine-induced liver tumors in mice and chemoprevention by leuprorelin. *Jpn J Cancer Res* **92**:249–256.

Naugler, W.E., Sakurai, T., Kim, S., Maeda, S., Kim, K., Elsharkawy, A.M., and Karin, M. 2007. Gender disparity in liver cancer due to sex differences in MyD88-dependent IL-6 production. *Science* **317**:121–124.

Okayasu, I., Ohkusa, T., Kajiura, K., Kanno, J., and Sakamoto, S. 1996. Promotion of colorectal neoplasia in experimental murine ulcerative colitis. *Gut* **39**:87–92.

Oshima, M., Dinchuk, J.E., Kargman, S.L., Oshima, H., Hancock, B., Kwong, E., Trzaskos, J.M., Evans, J.F., and Taketo, M.M. 1996. Suppression of intestinal polyposis in Apc delta716 knockout mice by inhibition of cyclo-oxygenase 2 (COX-2). *Cell* **87**:803–809.

Pham, C.G., Bubici, C., Zazzeroni, F., Papa, S., Jones, J., Alvarez, K., Jayawardena, S., De Smaele, E., Cong, R., Beaumont, C., Torti, F.M., Torti, S.V., Franzoso, G. 2004. Ferritin heavy chain upregulation by NF-kappaB inhibits TNFalpha-induced apoptosis by suppressing reactive oxygen species. *Cell* **119**:529–542.

Pollard, J.W. 2004. Tumour-educated macrophages promote tumour progression and metastasis. *Nat Rev Cancer* **4**:71–78.

Roder, D.M. 2002. The epidemiology of gastric cancer. *Gastric Cancer* **5** (Suppl. 1):5–11.

Rothwarf, D.M. and Karin, M. 1999. The NF-kappa B activation pathway: a paradigm in information transfer from membrane to nucleus. *Sci STKE* 26 Oct 1999. **1999**(5):p. re1. [DOI: 10.1126/stke.1999.Sre1].

Sakurai, T., He, G., Matsuzawa, A., Yu, G.Y., Maeda, S., Hardiman, G., Karin, M. (2008) Hepatocyte necrosis induced by oxidative stress and IL-1α release mediate carcinogen-induced compensatory proliferation and liver tumori-genesis. *Cancer Cell* **14**:156–165.

Sakurai, T., Maeda, S., Chang, L., and Karin, M. 2006. Loss of hepatic NF-kB activity enhances chemical hepatocarcinogenesis through sustained c-Jun N-terminal kinase 1 activation. *Proc Natl Acad Sci U S A* **103**: 10544–10551.

Sarma, D.S., Rao, P.M., and Rajalakshmi, S. 1986. Liver tumour promotion by chemicals: models and mechanisms. *Cancer Surv* **5**:781–798.

Sil, A.K., Maeda, S., Sano, Y., Roop, D.R., and Karin, M. 2004. IkB kinase-a acts in the epidermis to control skeletal and craniofacial morphogenesis. *Nature* **428**:660–664.

Steinbach, G., Lynch, P.M., Phillips, R.K., Wallace, M.H., Hawk, E., Gordon, G.B., Wakabayashi, N., Saunders, B., Shen, Y., Fujimura, T., Su, L.K., Levin, B. 2000. The effect of celecoxib, a cyclooxygenase-2 inhibitor, in familial adenomatous polyposis. *N Engl J Med* **342**:1946–1952.

Tergaonkar, V., Pando, M., Vafa, O., Wahl, G., and Verma, I. 2002. p53 stabi-lization is decreased upon NFkappaB activation: a role for NFkappaB in acquisition of resistance to chemotherapy. *Cancer Cell* **1**:493–503.

Thorgeirsson, S.S. and Grisham, J.W. 2002. Molecular pathogenesis of human hepatocellular carcinoma. *Nat Genet* **31**:339–346.

Zong, W.X. and Thompson, C.B. 2006. Necrotic death as a cell fate. *Genes Dev* **20**:1–15.

Zou, Z., Anisowicz, A., Hendrix, M.J., Thor, A., Neveu, M., Sheng, S., Rafidi, K., Seftor, E., and Sager, R. 1994. Maspin, a serpin with tumor-suppressing activity in human mammary epithelial cells. *Science* **263**:526–529.

FORMER OFFICERS OF THE HARVEY SOCIETY

1905–1906

President: GRAHAM LUSK
Vice-President: SIMON FLEXNER
Treasurer: FREDERIC S. LEE
Secretary: GEORGE B. WALLACE

Council:
C.A. HERTER
S.J. MELTZER
EDWARD K. DUNHAM

1906–1907

President: GRAHAM LUSK
Vice-President: SIMON FLEXNER
Treasurer: FREDERIC S. LEE
Secretary: GEORGE B. WALLACE

Council:
C.A. HERTER
S.J. MELTZER
JAMES EWING

1907–1908

President: GRAHAM LUSK
Vice-President: JAMES EWING
Treasurer: EDWARD K. DUNHAM
Secretary: GEORGE B. WALLACE

Council:
SIMON FLEXNER
THEO C. JANEWAY
PHILIP H. HISS, JR.

1908–1909

President: JAMES EWING
Vice-President: SIMON FLEXNER
Treasurer: EDWARD K. DUNHAM
Secretary: FRANCIS C. WOOD

Council:
GRAHAM LUSK
S.J. MELTZER
ADOLPH MEYER

1909–1910*

President: JAMES EWING
Vice-President: THEO C. JANEWAY
Treasurer: EDWARD K. DUNHAM
Secretary: FRANCIS C. WOOD

Council:
GRAHAM LUSK
S.J. MELTZER
W.J. GIES

1910–1911

President: SIMON FLEXNER
Vice-President: JOHN HOWLAND
Treasurer: EDWARD K. DUNHAM
Secretary: HAVEN EMERSON

Council:
GRAHAM LUSK
S.J. MELTZER
JAMES EWING

---

*At the Annual Meeting of May 18, 1909, these officers were elected. In publishing the 1909–1910 volume their names were omitted, possibly because in that volume the custom of publishing the names of the incumbents of the current year was changed to publishing the names of the officers selected for the ensuing year.

1911–1912

President: S.J. MELTZER
Vice-President: FREDERIC S. LEE
Treasurer: EDWARD K. DUNHAM
Secretary: HAVEN EMERSON

Council:
    GRAHAM LUSK
    JAMES EWING
    SIMON FLEXNER

1912–1913

President: FREDERIC S. LEE
Vice-President: WM. H. PARK
Treasurer: EDWARD K. DUNHAM
Secretary: HAVEN EMERSON

Council:
    GRAHAM LUSK
    S.J. MELTZER
    WM. G. MACCALLUM

1913–1914

President: FREDERIC S. LEE
Vice-President: WM. G. MACCALLUM
Treasurer: EDWARD K. DUNHAM
Secretary: AUGUSTUS B. WADSWORTH

Council:
    GRAHAM LUSK
    WM. H. PARK
    GEORGE B. WALLACE

1914–1915

President: WM. G. MACCALLUM
Vice-President: RUFUS I. COLE
Treasurer: EDWARD K. DUNHAM
Secretary: JOHN A. MANDEL

Council:
    GRAHAM LUSK
    FREDERIC S. LEE
    W.T. LONGCPE

1915–1916*

President: GEORGE B. WALLACE
Treasurer: EDWARD K. DUNHAM
Secretary: ROBERT A. LAMBERT

Council:
    GRAHAM LUSK
    RUFUS I. COLE
    NELLIS B. FOSTER

1916–1917

President: GEORGE B. WALLACE
Vice-President: RUFUS I. COLE
Treasurer: EDWARD K. DUNHAM
Secretary: ROBERT A. LAMBERT

Council:
    GRAHAM LUSK[†]
    W.T. LONGCOPE
    S.R. BENEDICT
    HANS ZINSSER

1917–1918

President: EDWARD K. DUNHAM
Vice-President: RUFUS I. COLE
Treasurer: F.H. PIKE
Secretary: A.M. PAPPENHEIMER

Council:
    GRAHAM LUSK
    GEORGE B. WALLACE
    FREDERIC S. LEE
    PEYTON ROUS

---

*Dr. William G. MacCallum resigned after election. On Doctor Lusk's motion Doctor George B. Wallace was made President – no Vice President was appointed.

[†]Doctor Lusk was made Honorary permanent Counsellor.

1918–1919

President: GRAHAM LUSK
Vice-President: RUFUS I. COLE
Treasurer: F.H. PIKE
Secretary: K.M. VOGEL

Council:
GRAHAM LUSK
JAMES W. JOBLING
FREDERIC S. LEE
JOHN AUER

1919–1920

President: WARFIELD T. LONGCOPE
Vice-President: S.R. BENEDICT
Treasurer: E.H. PIKE
Secretary: K.M. VOGEL

Council:
GRAHAM LUSK
HANS ZINSSER
FREDERIC S. LEE
GEORGE B. WALLACE

1920–1921*

President: WARFIELD T. LONGCOPE
Vice-President: S.R. BENEDICT
Treasurer: A.M. PAPPENHEIMER
Secretary: HOMER F. SWIFT

Council:
GRAHAM LUSK
FREDERIC S. LEE
HANS ZINSSER
GEORGE B. WALLACE

1921–1922

President: RUFUS I. COLE
Vice-President: S.R. BENEDICT
Treasurer: A.M. PAPPENHEIMER
Secretary: HOMER F. SWIFT

Council:
GRAHAM LUSK
HANS ZINSSER
H.C. JACKSON
W.T. LONGCOPE

1922–1923

President: RUFUS I. COLE
Vice-President: HANS ZINSSER
Treasurer: CHARLES C. LIEB
Secretary: HOMER F. SWIFT

Council:
GRAHAM LUSK
W.T. LONGCOPE
H.C. JACKSON
S.R. BENEDICT

1923–1924

President: EUGENE F. DUBOIS
Vice-President: HOMER F. SWIFT
Treasurer: CHARLES C. LIEB
Secretary: GEORGE M. MACKENZIE

Council:
GRAHAM LUSK
ALPHONSE R. DOCHEZ
DAVID MARINE
PEYTON ROUS

---

*These officers were elected at the Annual Meeting of May 21, 1920 but were omitted in the publication of the 1919–20 volume.

1924–1925

President: EUGENE F. DUBOIS
Vice-President: PEYTON ROUS
Treasurer: CHARLES C. LIEB
Secretary: GEORGE M. MACKENZIE

Council:
  GRAHAM LUSK
  RUFUS COLE
  HAVEN EMERSON
  WM. H. PARK

1925–1926

President: HOMER F. SWIFT
Vice-President: H.B. WILLIAMS
Treasurer: HAVEN EMERSON
Secretary: GEORGE M. MACKENZIE

Council:
  GRAHAM LUSK
  EUGENE F. DUBOIS
  WALTER W. PALMER
  H.D. SENIOR

1926–1927

President: WALTER W. PALMER
Vice-President: WM. H. PARK
Treasurer: HAVEN EMERSON
Secretary: GEORGE M. MACKENZIE

Council:
  GRAHAM LUSK
  HOMER F. SWIFT
  A.R. DOCHEZ
  ROBERT CHAMBERS

1927–1928

President: DONAD D. VAN SLYKE
Vice-President: JAMES W. JOBLING
Treasurer: HAVEN EMERSON
Secretary: CARL A.L. BINGER

Council:
  GRAHAM LUSK
  RUSSEL L. CECIL
  WARD J. MACNEAL
  DAVID MARINE

1928–1929

President: PEYTON ROUS
Vice-President: HORATIO B. WILLIAMS
Treasurer: HAVEN EMERSON
Secretary: PHILIP D. MCMASTER

Council:
  GRAHAM LUSK
  ROBERT CHAMBERS
  ALFRED F. HESS
  H.D. SENIOR

1929–1930

President: G. CNABY ROBINSON
Vice-President: ALFRED F. HESS
Treasurer: HAVEN EMERSON
Secretary: DAYTON J. EDWARDS

Council:
  GRAHAM LUSK
  ALFRED E. COHN
  A.M. PAPPENHEIMER
  H.D. SENIOR

1930–1931

President: ALFRED E. COHN
Vice-President: J.G. HOPKINS
Treasurer: HAVEN EMERSON
Secretary: DAYTON J. EDWARDS

Council:
  GRAHAM LUSK
  O.T. AVERY
  A.M. PAPPENHEIMER
  S.R. DETWILER

1931–1932

President: J.W. JOBLING
Vice-President: HOMER W. SMITH
Treasurer: HAVEN EMERSON
Secretary: DAYTON J. EDWARDS

Council:
  GRAHAM LUSK
  S.R. DETWILER
  THOMAS M. RIVERS
  RANDOLPH WEST

1932–1933

President: ALFRED F. HESS
Vice-President: HAVEN EMERSON
Treasurer: THOMAS M. RIVERS
Secretary: EDGAR STILLMAN

Council:
  GRAHAM LUSK
  HANS T. CLARKE
  WALTER W. PALMER
  HOMER W. SMITH

1933–1934

President: ALFRED HESS*
Vice-President: ROBERT K. CANNAN
Treasurer: THOMAS M. RIVERS
Secretary: EDGAR STILLMAN

Council:
  STANLEY R. BENEDICT
  ROBERT F. LOEB
  WADE H. BROWN

1934–1935

President: ROBERT K. CANNAN
Vice-President: EUGENE L. OPIE
Treasurer: THOMAS M. RIVERS
Secretary: RANDOLPH H. WEST

Council:
  HERBERT S. GASSER
  B.S. OPPENHEIMER
  PHILIP E. SMITH

1935–1936

President: ROBERT K. CANNAN
Vice-President: EUGENE L. OPIE
Treasurer: THOMAS M. RIVERS
Secretary: RANDOLPH H. WEST

Council:
  ROBERT F. LOEB
  HOMER W. SMITH
  DAVID MARINE

---

*Dr. Hess died December 5, 1933.

1936–1937

President: EUGENE L. OPIE
Vice-President: PHILIP E. SMITH
Treasurer: THOMAS M. RIVERS
Secretary: MCKEEN CATTELL

Council:
    GEORGE B. WALLACE
    MARTIN H. DAWSON
    JAMES B. MURPHY

1937–1938

President: EUGENE L. OPIE
Vice-President: PHILIP E. SMITH
Treasurer: THOMAS M. RIVERS
Secretary: MCKEEN CATTELL

Council:
    GEORGE B. WALLACE
    MARTIN H. DAWSON
    HERBERT S. GASSER

1938–1939

President: PHILIP E. SMITH
Vice-President: HERBERT S. GASSER
Treasurer: KENNETH GOODNER
Secretary: MCKEEN CATTELL

Council:
    HANS T. CLARKE
    JAMES D. HARDY
    WILLIAM S. TILLETT

1939–1940

President: PHILIP E. SMITH
Vice-President: HERBERT S. GASSER
Treasurer: KENNETH
Secretary: THOMAS FRANCIS, JR.

Council:
    HANS T. CLARKE
    GOODNER N. CHANDLER FOOT
    WILLIAM S. TILLETT

1940–1941

President: HERBERT S. GASSER
Vice-President: HOMER W.
Treasurer: KENNETH GOODNER
Secretary: THOMAS FRANCIS, JR.

Council:
    SMITH N. CHANDLER FOOT
    VINCENTDU VIGNEAUD
    MICHAEL HEIDELBERGER

1941–1942

President: HERBERT S. GASSER
Vice-President: HOMER W. SMITH
Treasurer: KENNETH GOODNER
Secretary: JOSEPH C. HINSEY

Council:
    HARRY S. MUSTARD
    HAROLD G. WOLFF
    MICHAEL HEIDELBERGER

1942–1943

President: HANS T. CLARKE
Vice-President: THOMAS M. RIVERS
Treasurer: KENNETH GOODNER
Secretary: JOSEPH C. HINSEY

Council:
    ROBERT F. LOEB
    HAROLD G. WOLFF
    WILLIAM C. VON GLAHN

1943–1944

President: HANS T. CLARKE
Vice-President: THOMAS M. RIVERS
Treasurer: COLIN M. MACLEOD
Secretary: JOSEPH C. HINSEY

Council:
ROBERT F. LOEB
WILLIAM C. VON GLAHN
WADE W. OLIVER

1944–1945

President: ROBERT CHAMBERS
Vice-President: VINCENTDU VIGNEAUD
Treasurer: COLIN M. MACLEOD
Secretary: JOSEPH C. HINSEY

Council:
WADE W. OLIVER
MICHAEL HEIDELBERGER
PHILIP D. MCMASTER

1945–1946

President: ROBERT CHAMBERS
Vice-President: VINCENTDU VIGNEAUD
Treasurer: COLIN M. MACLEOD
Secretary: EDGAR G. MILLER, JR.

Council:
PHILIP D. MCMASTER
EARL T. ENGLE
FRED W. STEWART

1946–1947

President: VINCENTDU VIGNEAUD
Vice-President: WADE W. OLIVER
Treasurer: COLIN M. MACLEOD
Secretary: EDGAR G. MILLER, JR.

Council:
EARL T. ENGLE
HAROLD G. WOLFF
L. EMMETT HOLT, JR.

1947–1948

President: VINCENTDU VIGNEAUD
Vice-President: WADE W. OLIVER
Treasurer: HARRY B. VAN
Secretary: MACLYN MCCARTY

Council:
PAUL KLEMPERER
DYKE L. EMMETT HOLT, JR.
HAROLD G. WOLFF

1948–1949

President: WADE W. OLIVER
Vice-President: ROBERT F. LOEB
Treasurer: HARRY B. VAN DYKE
Secretary: MACLYN MCCARTY

Council:
PAUL KLEMPERER
SEVERO OCHOA
HAROLD L. TEMPLE

1949–1950

President: WADE W. OLIVER
Vice-President: ROBERT F. LOEB
Treasurer: JAMES B. HAMILTON
Secretary: MACLYN MCCARTY

Council:
WILLIAM S. TILLETT
SEVERO OCHOA
HAROLD L. TEMPLE

1950–1951

President: ROBERT F. LOEB
Vice-President: MICHAEL HEIDELBERGER
Treasurer: JAMES B. HAMILTON
Secretary: LUDWIN W. EICHNA

Council:
  WILLIAMU S. TILLETT
  A.M. PAPPENHEIMER, JR.
  DAVID P. BARR

1951–1952

President: RENE J. DUBOIS
Vice-President: MICHAEL HEIDELBERGER
Treasurer: JAMES B. HAMILTON
Secretary: LUDWIN W. EICHNA

Council:
  DAVID P. BARR
  ROBERT F. PITTS
  A.M. PAPPENHEIMER, JR.

1952–1953

President: MICHAEL HEIDELBERGER
Vice-President: SEVERO OCHOA
Treasurer: CHANDLER McC. BROOKS
Secretary: HENRY D. LAUSON

Council:
  ROBERT F. PITTS
  JEAN OLIVER
  ALEXANDER B. GUTMAN

1953–1954

President: SEVERO OCHOA
Vice-President: DAVID P. BARR
Treasurer: CHANDLER McC. BROOKS
Secretary: HENRY D. LAUSON

Council:
  JEAN OLIVER
  ALEXANDER B. GUTMAN
  ROLLIN D. HOTCHKISS

1954–1955

President: DAVID P. BARR
Vice-President: COLIN M. MACLEOD
Treasurer: CHANDLER McC. BROOKS
Secretary: HENRY D. LAUSON

Council:
  ALEXANDER B. GUTMAN
  ROLLIN D. HOTCHKISS
  DAVID SHEMIN

1955–1956

President: COLIN M. MACLEOD
Vice-President: FRANK L. HORSFALL, JR.
Treasurer: CHANDLER McC. BROOKS
Secretary: RULON W. RAWSON

Council:
  ROLLIN D. HOTCHKISS
  DAVID SHEMIN
  ROBERT F. WATSON

1956–1957

President: FRANK L. HORSFALL, JR.
Vice-President: WILLIAM S. TILLETT
Treasurer: CHANDLER McC. BROOKS
Secretary: RULON W. RAWSON

Council:
  DAVID SHEMIN
  ROBERT F. WATSON
  ABRAHAM WHITE

1957–1958

President: WILLIAM S. TILLETT
Vice-President: ROLLIN D. HOTCHKISS
Treasurer: CHANDLER McC. BROOKS
Secretary: H. SHERWOOD LAWRENCE

Council:
ROBERT F. WATSON
ABRAHAM WHITE
JOHN V. TAGGART

1958–1959

President: ROLLIN D. HOTCHKISS
Vice-President: ANDRE COURNAND
Treasurer: CHANDLER McC. BROOKS
Secretary: H. SHERWOOD LAWRENCE

Council:
ABRAHAM WHITE
JOHN V. TAGGART
WALSH McDERMOTT

1959–1960

President: ANDRE COURNAND
Vice-President: ROBERT F. PITTS
Treasurer: EDWARD J. HEHRE WALSH
Secretary: H. SHERWOOD LAWRENCE

Council:
JOHN V. TAGGART
McDERMOTT
ROBERT F. FURCHGOTT

1960–1961

President: ROBERT F. PITTS
Vice-President: DICKINSON W. RICHARDS
Treasurer: EDWARD J. HEHRE
Secretary: ALEXANDER G.BEARN

Council:
WALSH McDERMOTT
ROBERT F. FURCHGOTT
LUDWIG W. EICHNA

1961–1962

President: DICKINSON W. RICHARDS
Vice-President: PAUL WEISS
Treasurer: I. HERBERT SCHEINBERG
Secretary: ALEXANDER G. BEARN

Council:
ROBERT F. FURCHGOTT
LUDWIG W. EICHNA
EFRAIM RACKER

1962–1963

President: PAUL WEISS
Vice-President: ALEXANDER B. GUTMAN
Treasurer: I. HERBERT SCHEINBERG
Secretary: ALEXANDER G. BEARN

Council:
LUDWIG W. EICHNA
EFRAIM RACKER
ROGER L. GREIF

1963–1964

President: ALEXANDER B. GUTMAN
Vice-President: EDWARD L. TATUM
Treasurer: SAUL J. FARBER
Secretary: ALEXANDER G. BEARN

Council:
EFRAIM RACKER
ROGER L. GREIF
IRAVING M. LONDON

1964–1965

President: EDWARD TATUM
Vice-President: CHANDLER McC. BROOKS
Treasurer: SAUL J. FARBER
Secretary: RALPH L. ENGLE, JR.

Council:
  ROGER L. GREIF
  LEWIS THOMAS
  IRVING M. LONDON

1965–1966

President: CHANDLER MC C. BROOKS
Vice-President: ABRAHAM WHITE
Treasurer: SAUL J. FARBER
Secretary: RALPH L. ENGLE, JR.

Council:
  IRVING M. LONDON
  LEWIS THOMAS
  GEORGE K. HIRST

1966–1967

President: ABRAHAM WHITE
Vice-President: RACHMIEL LEVINE
Treasurer: SAUL J. FARBER
Secretary: RALPH L. ENGLE, JR.

Council:
  LEWIS THOMAS
  GEORGE K. HIRST
  DAVID NACHMANSOHN

1967–1968

President: RACHMIEL LEVINE
Vice-President: SAUL J. FARBER
Treasurer: PAUL A. MARKS
Secretary: RALPH L. ENGLE, JR.

Council:
  GEORGE K. HIRST
  DAVID NACHMANSOHN
  MARTIN SONENBERG

1968–1969

President: SAUL J. FARBER
Vice-President: JOHN V. TAGGART
Treasurer: PAUL A. MARKS
Secretary: ELLIOTT F. OSSERMAN

Council:
  DAVID NACHMANSOHN
  MARTIN SONENBERG
  HOWARD A. EDER

1969–1970

President: JOHN V. TAGGART
Vice-President: BERNARD L. HORECKER
Treasurer: PAUL A. MARKS
Secretary: ELLIOTT F. OSSERMAN

Council:
  MARTIN SONENBERG
  HOWARD A. EDER
  SAUL J. FARBER

1970–1971

President: BERNARD L. HORECKER
Vice-President: MACLYN McCARTY
Treasurer: EDWARD C. FRANKLIN
Secretary: ELLIOTT F. OSSERMAN

Council:
  HOWARD A. EDER
  SAUL J. FARBER
  SOLOMON A. BERSON

1971–1972

President: MACLYN MCCARTY
Vice-President: ALEXANDER G. BEARN
Treasurer: EDWARD C. FRANKLIN
Secretary: ELLIOTT F. OSSERMAN

Council:
　SAUL J. FARBER
　SOLOMON A. BERSON
　HARRY EAGLE

1972–1973

President: ALEXANDER G. BEARN
Vice-President: PAUL A. MARKS
Treasurer: EDWARD C. FRANKLIN
Secretary: JOHN ZABRISKIE

Council:
　HARRY EAGLE
　JERARD HURWITZ

1973–1974

President: PAUL A. MARKS
Vice-President: IGOR TAMN
Treasurer: EDWARD C. FRANKLIN
Secretary: JOHN B. ZABRISKIE

Council:
　HARRY EAGLE
　CHARLOTTE FRIEND
　JERARD HURWITZ

1974–1975

President: IGOR TAMM
Vice-President: GERALD M. EDELMAN
Treasurer: STEPHEN I. MORSE
Secretary: JOHN B. ZABRISKIE

Council:
　JERARD HURWITZ
　H. SHERWOOD LAWRENCE
　CHARLOTTE FRIEND

1975–1976

President: GERALD M. EDELMAN
Vice-President: ELVIN A. KABAT
Treasurer: STEPHEN I. MORSE
Secretary: JOHN B. ZABRISKIE

Council:
　PAUL A. MARKS
　H. SHERWOOD LAWRENCE
　CHARLOTTE FRIEND

1976–1977

President: ELVIN A. KABAT
Vice-President: FREDPLUM
Treasurer: STEPHEN I. MORSE
Secretary: DONALD M. MARCUS

Council:
　H. SHERWOOD LAWRENCE
　PAUL A. MARKS
　BRUCE CUNNINGHAM

1977–1978

President: FRED PLUM
Vice-President: CHARLOTTE FRIEND
Treasurer: STEPHEN I. MORSE
Secretary: DONALD M. MARCUS

Council:
　PAUL A. MARKS
　BRUCE CUNNINGHAM
　VITTORIO DEFENDI

1978–1979

President: CHARLOTTE FRIEND
Vice-President: MARTIN SONENBERG
Treasurer: ALFRED STRACHER
Secretary: DONALD M. MARCUS

Council:
  BRUCE CUNNINGHAM
  VITTORIO DEFENDI
  DEWITT S. GOODMAN

1979–1980

President: MARTIN SONENBERG
Vice-President: KURT HIRSCHHORN
Treasurer: ALFRED STRACHER
Secretary: EMIL C. GOTSCHLICH

Council:
  VITTORIO DEFENDI
  DEWITT S. GOODMAN
  ORA ROSEN

1980–1981

President: KURT HIRSCHHORN
Vice-President: GERALD WEISSMANN
Treasurer: ALFRED STRACHER
Secretary: EMIL C. GOTSCHLICH

Council:
  RALPH NACHMAN
  DEWITT S. GOODMAN
  ORA ROSEN

1981–1982

President: GERALD WEISSMANN
Vice-President: DEWITT S. GOODMAN
Treasurer: ALFRED STRACHER
Secretary: EMIL C. GOTSCHLICH

Council:
  KURT HIRSCHHORN
  RALPH L. NACHMAN
  ORA ROSEN

1982–1983

President: DE WITT S. GOODMAN
Vice-President: MATTHEW D. SCHARFF
Treasurer: ALFRED STRACHER
Secretary: EMIL C. GOTSCHLICH

Council:
  KURT HIRSCHHORN
  RALPH L. NACHMAN
  GERALD WEISSMANN

1983–1984

President: MATTHEW D. SCHARFF
Vice-President: HAROLD S. GINSBERG
Treasurer: RICHARD A. RIFKIND
Secretary: EMIL C. GOTSCHLICH

Council:
  KURT HIRSCHHORN
  GERALD WEISSMANN
  JAMES P. QUIGLEY

1984–1985

President: HAROLD S. GINSBERG
Vice-President: JAMES E. DARNELL
Treasurer: RICHARD A. RIFKIND
Secretary: ROBERT J. DESNICK

Council:
  JAMES P. QUIGLEY
  MATTHEW D. SCHARFF
  GERALD WEISSMANN

1985–1986

President: JAMES E. DARNELL
Vice-President: DAVID SABATINI
Treasurer: RICHARD A. RIFKIND
Secretary: ROBERT J. DESNICK

Council:
JAMES P. QUIGLEY
MATTHEW D. SCHARFF
HAROLD S. GINSBERG

1986–1987

President: DAVID SABATINI
Vice-President: DONALD A. FISCHMAN
Treasurer: RICHARD A. RIFKIND
Secretary: ROBERT J. DESNICK

Council:
JAMES E. DARNELL
HAROLD S. GINSBERG
MATTHEW D. SCHARFF

1987–1988

President: DONALD A. FISCHMAN
Vice-President: JONATHAN R. WARNER
Treasurer: RICHARD A. RIFKIND
Secretary: ROBERT J. DESNICK

Council:
JAMES E. DARNELL
HAROLD S. GINSBERG
WILLIAM MCALLISTER
DAVID SABATINI

1988–1989

President: JONATHAN R. WARNER
Vice-President: ISIDORE S. EDELMAN
Treasurer: JOSEPH R. BERTINO
Secretary: ROBERT J. DESNICK

Council:
JAMES E. DARNELL
DONALD A. FISCHMAN
WILLIAM MCALLISTER
DAVID SABATINI

1989–1990

President: ISIDORE S. EDELMAN
Vice-President: DAVID J. LUCK
Treasurer: JOSEPH R. BERTINO
Secretary: PETER PALESE

Council:
DONALD A. FISCHMAN
ROCHELLE HIRSCHHORN
WILLIAM MCALLISTER
JONATHAN R. WARNER

1990–1991

President: DAVID J. LUCK
Vice-President: KENNETH BERNS
Treasurer: JOSEPH R. BERTINO
Secretary: PETER PALESE WILLIAM

Council:
ISIDORE S. EDELMAN
ROCHELLE HIRSCHHORN
MCALLISTER
JONATHAN R. WARNER

1991–1992

President: KENNETH I. BERNS
Vice-President: JERARD HURWITZ
Treasurer: JOSEPH R. BERTINO
Secretary: PETER PALESE

Council:
ISIDORE S. EDELMAN
ROCHELLE HIRSCHHORN
SUSAN HORWITZ
M.A.Q. SIDDIQUI

1992–1993

President: JERARD HURWITZ
Vice-President: RUTH S. NUSSENZWEIG
Treasurer: JOSEPH R. BERTINO
Secretary: PETER PALESE

Council:
  KENNETH I. BERNS
  SUSAN HORWITZ
  M.A.Q. SIDDIQUI
  SAMUEL C. SILVERSTEIN

1993–1994

President: RUTH S. NUSSENZWEIG
Vice-President: ROBERT ROEDER
Treasurer: JOAN MASSAGUE
Secretary: PAUL B. LAZAROW

Council:
  SUSAN HORWITZ
  JERARD HURWITZ
  M.A.Q. SIDDIQUI
  SAMUEL C. SILVERSTEIN

1994–1995

President: ROBERT ROEDER
Vice-President: DAVID HIRSH
Treasurer: JOAN MASSAGUE
Secretary: PAUL B. LAZAROW

Council:
  EVA B. CRAMER
  LORRAINE GUDAS
  MARSHALL HORWITZ
  JERARD HURWITZ
  RUTH NUSSENZWEIG
  SAMUEL C. SILVERSTEIN

1995–1996

President: DAVID I. HIRSH
Vice-President: STUART AARONSON
Treasurer: SCOTT EMMONS
Secretary: PAUL B. LAZAROW

Council:
  EVA B. CRAMER
  LORRAINE GUDAS
  MARSHALL HORWITZ
  KENNETH MARIANS
  LENNART PHILIPSON
  ROBERT G. ROEDER

1996–1997

President: STUART AARONSON
Vice-President: LENNART RHILIPSON
Treasurer: SCOTT EMMONS
Secretary: PAUL B. LAZAROW

Council:
  EVA B. CRAMER
  LORRAINE J. GUDAS
  DAVID I. HIRSH
  MARSHALL HOROWITZ
  KENNETH MARIANS
  PETER MODEL

1997–1998

President: LENNART PHILIPSON
Vice-President: PETER MODEL
Vice-President: DINSHAW PATEL
Treasurer: SCOTT EMMONS
Secretary: PAUL B. LAZAROW

Council:
STUART AARONSON
DAVID I. HIRSH
FREDERICK R. MAXFIELD
CAROL PRIVES
RONALD F. RIEDER

1998–1999

President: DINSHAW PATEL
Vice-President: RAJU KUCHERLAPATI
Treasurer: SCOTT EMMONS
Secretary: SARIO L.C. WOO

Council:
MARION CARLSON
RUTH LEHMAN
FREDERICK R. MAXFIELD
PETER MODEL
CAROL PRIVES
RONALD F. RIEDER

1999–2000

President: RAJU KUCHERLAPATI
Vice-President: TITIADE LANGE
Treasurer: SAVIO L.C. WOO
Secretary: SCOTT W. EMMONS

Council:
MARION CARLSON
RUTH LEHMAN
FREDERICK R. MAXFIELD
CAROL PRIVES
RONALD F. RIEDER
MICHAEL YOUNG

2000–2001

President: TITIA DELANGE
Vice-President: CAROL PRIVES
Treasurer: SCOTT EMMONS
Secretary: SAVIO L.C. WOO

Council:
OLAF S. ANDERSON
ROBERT BENEZRA
MARIAN CARLSON
RUTH LEHMAN
ROBERT H. SINGER
MICHAEL W. YOUNG

2001–2002

President: CAROL PRIVES
Vice-President: FREDERICK MAXFIELD
Treasurer: MICHAEL W. YOUNG
Secretary: SAVIO L.C. WOO

Council:
OLAF SPARRE ANDERSON
ROBERT BENEZRA
MAGDA KONARSKA
DAN LITTMAN
CAROL MASON
ROBERT H. SINGER

2002–2003

President: FREDERICK MAXFIELD
Vice-President: PETER PALESE
Treasurer: MICHAEL W. YOUNG
Secretary: ROBERT H. SINGER

Council:
OLAF SPARRE ANDERSON
ROBERT BENEZRA
MARIE FILBIN
GORDON M.KELLER
MAGDA KONARSKA
DON LITTMAN
JOSEF MICHL
VIRGINIA E. PAPAIOANNOU
PAMELA STANLEY

2003–2004

President: PETER PALESE
Vice-President: KATHRYN ANDERSON
Treasurer: MICHAEL W. YOUNG
Secretary: ROBERT H. SINGER

Council:
MARIE FILBIN
GORDON M. KELLER
MAGDA KONARSKA
DAN LITTMAN
JOAN MASSAGUE
JOSEF MICHL
VIRGINIA E. PAPAIOANNOU
PAMELA STANLEY
HAREL WEINSTEIN

2004–2005

President: KATHRYN V. ANDERSON
Vice-President: DAN LITTMAN
Treasurer: ROBERT BENEZRA
Secretary: ROBERT H. SINGER

Council:
MARIE FILBIN
GORDON M. KELLER
DAN LITTMAN
JOAN MASSAGUE
JOSEF MICHL
VIRGINIA E. PAPAIOANNOU
DAVID RON
PAMELA STANLEY
HAREL WEINSTEIN

2005–2006

President: DAN LITTMAN
Vice-President: E. RICHARD STANLEY
Treasurer: ROBERT BENEZRA
Secretary: ROBERT H. SINGER

Council:
ANA MARIA CUERVO
MARIE FILBIN
ADOLFO GARCIA-SASTRE
RICHARD KESSIN
JOAN MASSAGUÉ
MICHEL NUSSENZWEIG
DAVID RON
M.A.Q. SIDDIQUI
HAREL WEINSTEIN

CUMULATIVE AUTHOR INDEX*

DR. STUART A. AARONSON, 1991–92 (h)
DR. JOHN J. ABEL, 1923–24 (d)
DR. JOHN ABELSON, 1989–90 (h)
PROF. J.D. ADAMI, 1906–07 (d)
DR. ROGER ADAMS, 1941–42 (d)
DR. THOMAS ADDIS, 1927–28 (d)
DR. JULIUS ADLER, 1976–77 (d)
DR. E.D. ADRIAN, 1931–32 (h)
DR. FULLER ALBRIGHT, 1942–43 (h)
DR. FRANZ ALEXANDER, 1930–31 (h)
DR. FREDERICK ALLEN, 1916–17 (d)
DR. FREDERICK W. ALT, 1992–93
DR. JOHN F. ANDERSON, 1908–09 (d)
DR. KATHRYN V. ANDERSON, 2006–07 (a)
DR. R.J. ANDERSON, 1939–40 (d)
DR. CHRISTOPHER H. ANDREWS, 1961–62 (h)
DR. CHRISTIAN B. ANFINSEN, 1965–66 (h)
PROF. G.V. ANREP, 1934–35 (h)
DR. CHARLES ARMSTRONG, 1940–41 (d)
DR. LUDWIG ASCHOFF, 1923–24 (d)
DR. LEON ASHER, 1922–23 (d)
DR. W.T. ASTBURY, 1950–51 (h)
DR. EDWIN ASTWOOD, 1944–45 (d)
DR. JOSEPH C. AUB, 1928–29 (d)
DR. K. FRANK AUSTEN, 1977–78 (h)
DR. RICHARD AXEL, 1983–84 (a)
DR. JULIUS AXELROD, 1971–72 (h)
DR. E.R. BALDWIN, 1914–15 (d)
DR. DAVID BALTIMORE, 1974–75 (h)
PROF. JOSEPH BARCROFT, 1921–22 (d)
DR. PHILIP BRAD, 1921–22 (d)
DR. CORNELIA I. BARGMANN, 2000–01
DR. H.A. BARKER, 1949–50 (h)
DR. LEWELLYS BARKER, 1905–06 (d)
DR. BONNIE L. BASSLER, 2004–05
DR. JULIUS BAUER, 1932–33 (d)
PROF. WILLIAM M. BAYLISS, 1921–22 (d)
DR. DAVID BEACH, 1990–91 (h)
DR. FRANK BEACH, 1947–48 (h)
DR. GEORGE W. BEALE, 1944–45 (h)

DR. ALEXANDER G. BEARN, 1974–75 (a)
DR. ALBERT BEHNKE, 1941–42 (h)
DR. BARUJ BENACERRAF, 1971–72 (a)
PROF F.G. BENEDICT, 1906–07 (d)
DR. STANLEY BENEDICT, 1915–16 (d)
DR. STEPHEN J. BENKOVIC, 1991–92 (h)
DR. D. BENNETT, 1978–79 (a)
DR. M.V.L. BENNETT, 1982–83 (h)
PROF. R.R. BENSLEY, 1914–15 (d)
DR. SEYMOUR BENZER, 1960–61 (h)
DR. PAUL BERG, 1971–72 (h)
DR. MAX BERGMANN, 1935–36 (d)
DR. SUNE BERGSTRÖM, 1974–75 (h)
DR. ROBERT W. BERLINER, 1958–59 (h)
DR. SOLOMAN A. BERSON, 1966–67 (d)
DR. MARCEL C. BESSIS, 1962–63 (h)
DR. C.H. BEST, 1940–41 (h)
DR. A. BEIDEL, 1923–24 (d)
DR. RUPERT E. BILLINGHAM, 1966–67 (h)
DR. RICHARD J. BING, 1954–55 (a)
DR. J. MICHAEL BISHOP, 1982–83 (h)
DR. JOHN J. BRITNER, 1946–47 (d)
DR. ELIZABETH H. BLACKBURN, 1990–91 (h)
PROF. FRNACIS G. BLAKE, 1934–35 (d)
DR. ALFRED BLALOCK, 1945–46 (d)
DR. GUNTER BLOBEL, 1980–81 (a)
DR. KONRAD BLOCH, 1952–53 (h)
DR. BARRY R. BLOOM, 1988–89 (a)
DR. WALTER R. BLOOR, 1923–24 (d)
DR. DAVID BODIAN, 1956–57 (h)
DR. WALTER F. BODMER, 1976–77 (h)
DR. JAMES BONNER, 1952–53 (h)
DR. JULES BORDET, 1920–21 (d)
DR. DAVID BOTSTEIN, 1986–87 (h)
DR. WILLIAM T. BOVIE, 1922–23 (d)
DR. EDWARD A. BOYSE, 1971–72, 1975–76 (h)
DR. STANLEY E. BRADLEY, 1959–60 (a)
DR. DANIEL BARNTON, 1981–82 (a)
DR. ARMIN C. BRAUN, 1960–61 (h)

---

*(h), honorary; (a), active; (d), deceased.

169

DR. EUGNEE BRAUNWALD, 1975–76 (h)
PROF. F. BREMER, (h)*
DR. RALPH L. BRINSTER, 1984–85 (h)
PROF. T.G. BRODIE, 1909–10 (d)
DR. DETLEV W. BRONK, 1933–34 (d)
DR. MARIANNE BRONNER-FRASER, 2003–04
DR. B. BROUWER, 1925–26 (d)
DR. DONALD D. BROWN, 1980–81 (a)
DR. MICHAEL S. BROWN, 1977–78 (h)
DR. PATRICK O. BROWN, 2001–02 (a)
DR. WADE H. BROWN, 1928–29 (d)
DR. JOHN M. BUCHANAN, 1959–60 (h)
DR. LINDA B. BUCK, 2005–06
DR. JOHN CAIRNS, 1970–71 (h)
PROF. A. CALMETTE, 1908–09 (d)
DR. MELVIN CALVIN, 1950–51
DR. PIETRO DE CAMILLI, 2004–05
PROF. WALTER B. CANNON, 1911–12 (d)
DR. LEWIS C. CANTLEY, 2004–05
PROF. A.J. CARLSON, 1915–16 (d)
DR. MARC G. CARON, (h)
DR. WILLIAM B. CASTLE, 1934–35 (d)
PROF. W.E. CASTLE, 1910–11 (d)
DR. THOMAS R. CECH, 1986–87 (h)
DR. CONSTANCE L. CEPKO, 2001–02 (a)
DR. I.L. CHAIKOFF, 1951–52 (d)
DR. MICHAEL J. CHAMBERLIN, 1992–93
DR. ROBERT CHAMBERS, 1926–27 (d)
DR. B. CHANCE, 1953–54 (h)
DR. JEAN-PIERRE CHANGEAUX, 1979–80 (h)
DR. CHARLES V. CHAPIN, 1913–14 (d)
DR. ERWIN CHARGAFF, 1956–57 (h)
DR. MERRILL W. CHASE, 1965–66 (a)
DR. JIANDONG CHEN, 1944–45
DR. ALAN M. CHESNEY, 1929–30 (d)
DR. WAJ YUI CHEUNG, 1983–84 (h)
PROF. HANS CHIARI, 1910–11 (d)
DR. C.M. CHILD, 1928–29 (d)
PROF. RSSELL H. CHIRTENDEN, 1911–12 (d)
DR. PURNELL W. CHOPPIN, 1984–85 (a)

PROF. HENRY A. CHRISTIAN, 1915–16 (d)
DR. W. MANSFIELD CLARK, 1933–34 (d)
DR. ALBERT CLAUDE, 1947–58 (h)
DR. SAMUEL W. CLAUSEN, 1942–43 (d)
DR. PHILIP P. COHEN, 1964–65 (h)
DR. STANLEY N. COHEN, 1978–79 (h)
DR. ZANVIL A. COHN, 1981–82 (a)
DR. ALFRED E. COHN, 1927–28 (d)
DR. EDWIN F. COHN, 1927–28, 1938–39 (d)
PROF. OTTO COHNHEIM, 1909–10 (d)
DR. RUFUS COLE, 1913–14, 1929–30 (d)
DR. FRANCIS S. COLLINS, 1990–91 (h)
DR. J.B. COLLIP, 1925–26 (d)
DR. EDGAR L. COLLIS, 1926–27 (d)
DR. JULIUS H. COMROE, JR., 1952–53 (h)
DR. JAMES B. CONANT, 1932–33 (h)
PROF. EDWIN G. CONKLIN, 1912–13 (d)
DR. JEROME W. CONN, 1966–67 (h)
DR. ALBERT H. COONS, 1957–58 (d)
DR. CARL F. CORI, 1927–28, 1945–46 (h)
DR. GERTY T. CORI, 1952–53 (d)
DR. GEORGE W. CORNER, 1932–33 (h)
DR. GEORGE C. COTZIAS, 1972–73 (d)
PROF. W.T. COUNCILMAN, 1906–07 (d)
DR. ANDRE COURNAND, 1950–51 (a)
DR. E.V. COWDRY, 1922–23 (d)
DR. NICHOLAS R. COZZABELLI, 1991–92 (h)
DR. LYMAN C. CRAIG, 1949–59 (d)
DR. GEORGE CRILE, 1907–08 (d)
DR. S.J. CROWE, 1931–32 (d)
DR. PEDRO CUATRECASAS, 1984–85 (h)
DR. HARVEY CUSHING, 1910–11, 1932–33 (d)
PROF. ARTHUR R. CUSHNY, 1910–11 (d)
SIR HENRY DALE, 1919–20, 1936–37 (h)
DR. I. DE BURGH DALY, 1935–36 (d)
DR. C.H. DANFORTH, 1938–39 (d)
DR. JAMES F. DANIELLI, 1962–63 (h)
DR. JAMES E. DARNELL, JR., 1973–74 (a)
DR. C.B. DAVENPORT, 1908–09 (d)

---

*Did not present lecture because of World War II.

Dr. Earl W. Davie, 1981–82 (h)
Dr. Bernard D. Davis, 1954–55 (h)
Dr. Mark M. Davis, 1999–00 (?)
Dr. Christian de Duve, 1963–64 (h)
Dr. Max Delbruck, 1945–46 (h)
Dr. Hector F. De Luca, 1979–80 (h)
Dr. F. D'Herelle, 1928–29 (d)
Dr. John H. Dingle, 1956–57 (d)
Dr. Frank J. Dixon, 1962–63 (h)
Dr. A.R. Dochez, 1924–25 (d)
Dr. Jane Dodd, 1990–91 (h)
Dr. E.C. Dodds, 1934–35 (h)
Dr. E.A. Doisy, 1933–34 (d)
Dr. Vincent P. Dole, 1971–72 (h)
Prof. Henry H. Donaldson, 1916–17 (d)
Dr. W. Ford Doolittle, 2003–04
Dr. Russell Dorer, 1994–95
Dr. Paul Doty, 1958–59 (h)
Prof. Georges Dreyer, 1919–20 (d)
Dr. Cecil K. Drinker, 1937–38 (d)
Dr. J.C. Drummond, 1932–33 (d)
Dr. Lewis I. Dublin, 1922–23 (h)
Dr. Eugene F. Du Bois, 1915–16, 1938–39, 1946–47 (d)
Dr. Rene J. Du Bois, 1939–40 (a)
Dr. Renato Dulbecco, 1967–68 (h)
Dr. E.K. Dunham, 1917–19 (d)
Dr. L.C. Dunn, 1939–40 (d)
Dr. Susan K. Dutcher, 2006–07 (a)
Dr. Vincent du Vigneaud, 1942–43, 1945–55 (a)
Dr. R.E. Dyer, 1933–34 (h)
Dr. Harry Eagle, 1959–60 (a)
Dr. E.M. East, 1930–31 (d)
Dr. J.C. Eccles, 1955–56 (h)
Dr. Gerald M. Edelman, 1972–73 (a)
Prof. R.S. Edgar, 1967–68 (h)
Dr. David L. Edsall, 1907–08 (d)
Dr. John T. Edsall, 1966–67 (h)
Dr. William Einthoven, 1924–25 (d)
Dr. Herman N. Eisen, 1964–65 (h)
Dr. Robert N. Eisenman, 2000–01
Dr. Joel Elkes, 1961–62 (h)
Dr. Stephen J. Elledge, 2001–02

Dr. C.A. Elvehjem, 1939–40 (d)
Dr. Haven Emerson, 1954–55 (d)
Dr. John F. Enders, 1947–48, 1963–64 (h)
Dr. Boris Ephrussi, 1950–51 (h)
Dr. Joseph Erlanger, 1912–13, 1926–27 (h)
Dr. Earl A. Evans, Jr., 1943–44 (h)
Dr. Herbert M. Evans, 1923–24 (h)
Dr. Ronald M. Evans, 1994–95
Dr. James Ewing, 1907–08 (d)
Dr. Knud Faber, 1925–26 (d)
Dr. Stanley Falkow, 1997–98
Dr. W. Falta, 1908–09 (d)
Dr. W.O. Fenn, 1927–28 (d)
Dr. Frank Fenner, 1956–57 (h)
Dr. Gerald R. Fink, 1988–89 (a)
Dr. H.O.L. Fischer, 1944–45 (d)
Dr. L.B. Flexner, 1951–52 (h)
Dr. Simon Flexner, 1911–12 (d)
Dr. Otto Folin, 1907–08, 1919–20 (d)
Dr. Judah Folkman, 1996–97
Prof. John A. Fordyce, 1914–15 (d)
Dr. Nellis B. Foster, 1920–21 (d)
Dr. Edward Francis, 1927–28 (d)
Dr. Thomas Francis, Jr., 1941–42 (d)
Dr. H. Fraenkel-Conrat, 1956–57 (h)
Dr. Robert T. Frank, 1930–31 (d)
Dr. Edward C. Franklin, 1981–82 (d)
Dr. Donald S. Frederickson, 1972–73 (h)
Dr. Irwin Fridovich, 1983–84 (h)
Dr. Carlotte Friend, 1976–77 (d)
Dr. C. Fromageot, 1953–56 (h)
Dr. Joseph S. Fruton, 1955–56 (a)
Dr. Elaine Fuchs, 1999–00 (h)
Dr. John F. Fulton, 1935–36 (d)
Dr. E.J. Furshpan, 1980–81 (a)
Dr. Jacob Furth, 1967–68 (a)
Dr. D. Carleton Gadjusek, 1976–77 (h)
Dr. Ernest F. Gale, 1955–56 (h)
Dr. Joseph G. Gall, 1975–76 (h)
Dr. T.F. Gallagher, 1956–57 (h)
Dr. Robert C. Gallo, 1983–84 (h)

Dr. James L. Gamble, 1946–47 (d)

Dr. Herbert S. Gasser, 1936–37 (d)

Dr. Frederick P. Gay, 1914–15, 1930–31 (d)

Dr. Walter J. Gehring, 1985–86 (h)

Dr. Eugene M.K. Geiling, 1941–42 (d)

Dr. Isidore Gersh, 1924–50 (h)

Dr. George O. Gey, 1954–55 (d)

Dr. John H. Gibbon, 1957–58 (d)

Dr. Alfred G. Gilman, 1989–90 (h)

Dr. David V. Goedell, 1996–97

Dr. Larry Gold, 1995–96

Dr. Harry Goldblatt, 1937–38 (h)

Dr. Joseph L. Goldstein, 1977–78 (h)

Dr. Robert A. Good, 1971–72 (a)

Dr. DeWirt S. Goodman, 1985–86 (a)

Dr. Earnest W. Goodpasture, 1929–30 (d)

Dr. Carl W. Gottschalk, 1962–63 (h)

Dr. J. Gough, 1957–58 (h)

Prof. J.I. Gowans, 1968–69 (h)

Dr. Evarts A. Graham, 1923–24, 1933–34 (d)

Dr. S. Granick, 1948–49 (h)

Dr. David E. Green, 1956–57 (h)

Dr. Howard Green, 1978–79 (h)

Dr. Michael R. Green, 1992–93

Dr. Michael E. Greenberg, 2006–07 (a)

Dr. Paul Greengard, 1979–80 (h)

Prof. R.A. Gregory, 1968–69 (h)

Dr. Carol W. Greider, 2000–01

Dr. Donald R. Griffin, 1975–76 (h)

Dr. Jerome Gross, 1972–73 (h)

Dr. Roger Guillemin, 1975–76 (h)

Dr. I.C. Gunsalus, 1949–50 (h)

Dr. John B. Gurdon, 1973–74 (h)

Dr. Christine Guthrie, 1994–95

Dr. Alexander B. Gutman, 1964–65 (a)

Dr. J.S. Haldane, 1916–17 (d)

Dr. William S. Haisted, 1913–14 (d)

Dr. H.J. Hamburger, 1922–23 (d)

Dr. Hidesaburo Hanafusa, 1979–80 (a)

Dr. J.D. Hardy, 1953–54 (d)

Sir William Hardy, 1930–31 (d)

Dr. Harry Harris, 1980–81 (a)

Prof. Henry Harris, 1969–70 (h)

Dr. Ross G. Harrison, 1907–08, 1933–34 (d)

Dr. Stephen C. Harrison, 1989–90 (h)

Dr. H.K. Hartline, 1941–42 (h)

Dr. Leonard H. Hartwell, 1994–95

Dr. E. Newton Harvey, 1944–45 (h)

Dr. A. Baird Hastings, 1940–41 (a)

Dr. Selig Hecht, 1937–38 (d)

Prof. Sven H. Hedin, 1913–14 (d)

Dr. Michael Heldelberger, 1932–33 (a)

Prof. Ludvig Hektoen, 1909–10 (d)

Prof. L.J. Henderson, 1914–15 (d)

Dr. Yandell Henderson, 1917–18 (d)

Dr. Wayne A. Hendrickson, 1992–93

Dr. James B. Herrick, 1930–31 (d)

Dr. A.D. Hershey, 1955–56 (h)

Dr. Ira Herskowitz, 1985–86 (h)

Prof. Christian Herter, 1906–07 (d)

Dr. Alfred F. Hess, 1920–21 (d)

Dr. A.V. Hill, 1924–25 (h)

Dr. Bertil Hille, 1986–87 (h)

Dr. George Hirst, 1948–49 (a)

Dr. Philip H. Hiss, 1908–09 (d)

Dr. Dorothy C. Hodgkin, 1965–66 (h)

Dr. Alan F. Hofmann, 1978–79 (h)

Dr. Kiaus Hofmann, 1963–64 (h)

Dr. Brigid L.M. Hogan, 1996–97

Dr. Leroy Hood, 1996–97

Prof. F. Gowland Hopkins, 1920–21 (d)

Dr. Bernand L. Horecker, 1961–62 (a)

Dr. Frank Horsfall, Jr., 1952–53 (d)

Dr. H. Robert Horvitz, 1988–89 (a)

Dr. R.D. Hotchkiss, 1953–54 (a)

Dr. B.A. Houssay, 1935–36 (h)

Prof. W.H. Howell, 1905–06, 1916–17 (d)

Dr. John Holand, 1912–13, 1922–23 (d)

Dr. David H. Hubel, 1976–77 (h)

Prof. G. Carl Huber, 1909–10 (d)

Dr. A.J. Hudspeth, 2001–02

Dr. Robert J. Huebner, 1960–61 (h)

Dr. Charles Huggins, 1946–47 (h)
Dr. David M. Hume, 1968–69 (d)
Dr. Tony Hunter, 1999–00 (h)
Prof. George Huntington, 1906–07 (d)
Dr. Jerard Hurwitz, 1968–69 (a)
Dr. Hugh Huxley, 1964–65 (h)
Dr. Richard O. Hynes, 1985–86 (h)
Dr. Vernon M. Ingram, 1965–66 (h)
Dr. Kurt J. Isselbacher, 1973–74 (h)
Dr. A.C. Ivy, 1931–32 (d)
Dr. Tyler Jacks, 2005–06
Dr. Francois Jacob, 1959–60 (h)
Dr. Merkel Jacobs, 1926–27 (d)
Dr. Walter A. Jacobs, 1923–24 (d)
Dr. Rudolph Jaenisch, 2002–03 (a)
Dr. Lily Yeh Jan, 1999–00
Dr. Yuh Nung Jan, 1999–00 (a)
Prof. Theodore C. Janeway, 1912–13 (d)
Dr. Joseph Jastro, 1907–08 (d)
Prof. H.S. Jennings, 1911–12 (d)
Dr. Niels K. Jerne, 1974–75 (h)
Dr. Thomas M. Jessell, 1990–91
Dr. J.W. Jobling, 1916–17 (d)
Dr. Edwin O. Jordan, 1907–08 (d)
Prof. Elijott P. Joslin, 1914–15 (d)
Dr. Alfred Jost, 1958–59 (h)
Dr. David Julius, 2005–06
Dr. Elvin A. Kabat, 1950–51 (a)
Dr. H. Ronald Kaback, 1987–88 (a)
Prof. Herman M. Kalckar, 1949–50 (h)
Dr. Yuet Wai Kan, 1980–81 (h)
Dr. Eric R. Kandel, 1977–78 (a)
Dr. Henry S. Kaplan, 1968–69 (h)
Dr. Nathan O. Kaplan, 1970–71 (h)
Dr. Michael Karin, 2006–07 (a)
Dr. Arthur Karlin, 1989–90 (a)
Dr. Ephraim Katchalski, 1963–64 (h)
Dr. Thomas J. Kelly, 1989–90 (h)
Prof. E.C. Kendall, 1919–20 (h)
Dr. Eugene P. Kennedy, 1961–62 (h)
Dr. Cynthia Kenyon, 2004–05
Dr. Seymour S. Kety, 1975–76 (h)
Dr. H. Gobind Khorana, 1966–67 (h)

Dr. Edwin D. Kilbourne, 1977–78 (a)
Dr. Peter S. Kim, 2001–02
Dr. K. Kindstrom-Lang, 1938–39 (d)
Dr. Marc W. Kirschner, 1987–88 (a)
Dr. Richard D. Klausner, 1996–97
Dr. Nancy Kleckner, 1995–96
Dr. George Klein, 1973–74 (h)
Dr. P. Klemperer, 1953–54 (d)
Dr. A. Klug, 1978–79 (h)
Dr. B.C.J.G. Knight, 1947–48 (h)
Prof. Franz Knoop, 1912–13 (d)
Dr. F.C. Koch, 1937–38 (d)
Prof. W. Kolle, 1924–25 (d)
Dr. Richard D. Kolodner, 2001–02
Dr. Masakazu Konishi, 1990–91 (h)
Dr. Hilary Koprowski, 1964–65 (h)
Dr. Arthur Kornberg, 1957–58 (a)
Dr. Daniel E. Koshland, Jr., 1969–70 (h)
Prof. Albrecht Kossel, 1911–12 (d)
Dr. Allen K. Krause, 1921–22 (d)
Dr. H.A. Krebs, 1948–49 (h)
Dr. August Krogh, 1922–23 (d)
Dr. Stephen W. Kuffler, 1959–69 (h)
Dr. Henry G. Kunkel, 1963–64 (a)
Dr. L.O. Kunkel, 1932–33 (d)
Dr. Rebecca C. Lancefield, 1940–41 (a)
Dr. Eric Lander, 1997–98
Dr. Eugene M. Landis, 1936–37 (h)
Dr. Ernst Laquer, 1947–46 (d)
Dr. Henry A. Lardy, 1964–65 (h)
Dr. K.S. Lashley, 1930–31 (d)
Dr. H. Sherwood Lawrence, 1972–73 (a)
Dr. H.A. Lawson, 1927–28 (d)
Dr. J.B. Leathes, 1908–09 (d)
Dr. Philip Leder, 1978–79 (a)
Dr. Joshua Lederberg, 1957–58 (h)
Dr. Nicole M. Le Douarin, 1984–85 (h)
Dr. Frederic S. Lee, 1905–06, 1917–19 (d)
Dr. Robert J. Lefkowttz, 1990–91 (h)
Dr. W.E. LeGros Clark, 1962–63 (h)

Dr. Ruth Lehmann, 2005–06
Dr. A.L. Lehninger, 1953–54 (h)
Dr. Luis F. Leloir, 1960–61 (h)
Dr. Richard A. Lerner, 1996–97
Dr. Henry A. Lester, 1995–96
Dr. C. Levaditi, 1928–29 (d)
Dr. P.A. Levene, 1906–06 (d)
Dr. Rita Levi-Montalcini, 1964–65 (h)
Dr. Arnold J. Levine, 1994–95
Dr. Beth Levine, 2003–04
Dr. Sam Z. Levine, 1946–47 (d)
Dr. Howard B. Lewis, 1940–41 (d)
Dr. Paul A. Lewis, 1916–17 (d)
Prof. Thomas Lewis, 1914–15 (d)
Dr. Warren H. Lewis, 1925–26, 1935–36 (d)
Dr. Richard C. Lewontin, 1974–75 (h)
Dr. Chon Hao Li, 1950–51 (h)
Dr. Jiayuh Lin, 1994–95
Dr. Susan L. Lindquist, 2002–03 (a)
Dr. Karl P. Link, 1943–44 (h)
Dr. Frttz Lipmann, 1948–49 (d)
Dr. C.C. Little, 1921–22 (d)
Dr. Richard P. Lifton, 2004–05
Dr. Harvey Lodish, 1986–87 (h)
Prof. Jacques Loeb, 1910–11, 1920–21 (d)
Dr. Leo Loeb, 1940–41 (d)
Dr. Robert F. Loeb, 1941–42 (h)
Prof. A.S. Loevenhart, 1914–15 (d)
Dr. Otto Loewi, 1932–33 (d)
Dr. E.S. London, 1927–28 (h)
Dr. Irving M. London, 1960–61 (a)
Dr. C.N.H. Long, 1936–37 (h)
Dr. Esmond R. Long, 1929–30 (h)
Prof. Warfield T. Longcope, 1915–16 (d)
Dr. Pafael Lorente de No, 1946–47 (h)
Prof. Konrad Lorenz, 1959–60 (h)
Dr. Ricahrd Losick, 1994–95
Dr. William D. Lotspeich, 1960–61 (d)
Dr. Oliver H. Lowry, 1962–63 (a)
Dr. Hua Lu, 1994–95

Dr. Einar Lundsgaard, 1937–38 (d)
Dr. S.E. Luria, 1964–65 (h)
Dr. Graham Lusk, 1908–09, 1929–30 (d)
Dr. Andre Lwoff, 1945–55 (h)
Dr. Feodor Lynen, 1952–53 (h)
Dr. A.B. Macallum, 1908–09 (d)
Dr. W.G. MacCallum, 1908–09 (d)
Prof. J.J.R. MacLeod, 1913–1914 (d)
Dr. William deB. MacNider, 1928–29 (d)
Dr. Thorvald Madsen, 1924–25, 1936–37 (d)
Prof. A. Magnus-Levy, 1909–10 (d)
Dr. H.W. Magoun, 1951–52 (h)
Prof. Philip W. Majerus, 1986–87 (h)
Dr. E.B. Mallory, 1912–13 (d)
Dr. Thomas P. Maniatis, 1986–87 (h)
Dr. Frank C. Mann, 1927–28 (d)
Dr. E. Margollash, 1970–71 (h)
Dr. David Marine, 1923–24 (d)
Dr. Clement L. Markert, 1963–64 (h)
Dr. Andrew R. Marks, 2005–06
Dr. Paul A. Marks, 1970–71 (a)
Dr. Philippa Marrack, 1993–94
Dr. Guy Marrian, 1938–39 (h)
Prof. W. Mc Kim Marriott, 1919–20 (d)
Dr. E.K. Marshall, Jr., 1929–30 (d)
Dr. Diane Mathis, 2003–04
Dr. Brian W. Matthews, 1986–86 (h)
Dr. Manfred M. Mayer, 1976–77 (h)
Dr. Daniel Mazia, 1957–58 (h)
Dr. Maclyn McCarty, 1969–70 (a)
Prof. E.V. McCollum, 1916–17 (d)
Dr. Waish McDermott, 1967–68 (h)
Dr. Harden M. McDonnell (h)
Dr. W.D. McElroy, 1955–56 (h)
Dr. Steven Lanier McKnight, 1991–92 (h)
Dr. Philip D. McMaster, 1941–42 (h)
Dr. P.B. Medawar, 1956–57 (h)
Dr. Douglas Melton, 1997–98
Dr. Walter J. Meek, 1940–41 (d)
Prof. Alton Meister, 1967–68 (h)

Dr. John J. Mekalanos, 1993–94
Dr. S.J. Meltzer, 1906–07 (d)
Prof. LaFayette B. Mendel, 1905–06, 1914–15 (d)
Dr. R. Bruce Merrifield, 1971–72 (h)
Dr. Henry Metzger, 1984–85 (h)
Prof. Adolph Meyer, 1909–10 (d)
Prof. Hans Meyer, 1905–06 (d)
Dr. Karl Meyer, 1955–56 (h)
Dr. K.F. Meyer, 1939–40 (d)
Dr. Otto Meyerhof, 1922–23 (d)
Dr. Elliot M. Meyerowitz, 2000–01
Dr. Leonor Michaelis, 1926–27 (d)
Dr. William S. Miller, 1924–25 (d)
Prof. Charles S. Minot, 1905–06 (d)
Dr. George R. Minot, 1927–28 (d)
Dr. Beatrice Mintz, 1975–76 (h)
Dr. A.E. Mirsky, 1950–51 (h)
Dr. Timothy Mitchison, 2002–03 (a)
Dr. Jacques Monod, 1961–62 (h)
Dr. Carl V. Moore, 1958–59 (h)
Dr. Francis D. Moore, 1956–57 (h)
Dr. Stanford Moore, 1956–57 (h)
Prof. T.H. Morgan, 1905–06 (d)
Dr. Giuseppe Moruzzi, 1962–63 (h)
Dr. J. Howard Mueller, 1943–44 (d)
Prof. Friedrich Muller, 1906–07 (d)
Dr. H.J. Muller, 1947–48 (d)
Dr. Hans Muller-Eberhard, 1970–71 (a)
Prof. John R. Murlin, 1916–17 (d)
Dr. W.P. Murphy, 1927–28 (d)
Dr. David Nachmansohn, 1953–54 (h)
Dr. F.R. Nager, 1925–26 (d)
Dr. Lily Yeh Nan, 1999–00 (a)
Dr. Kim Nasmyth, 1992–93
Dr. Daniel Nathans, 1974–75 (h)
Dr. Stanley G. Nathenson, 1993–94 (a)
Dr. James V. Neel, 1960–61 (h)
Dr. Elizabeth F. Neufeld, 1979–80 (h)
Dr. Fred Neufeld, 1926–27 (d)
Sir Arthur Newsholme, 1920–21 (d)
Dr. Marshall W. Nirenberg, 1963–64 (h)

Dr. Hideyo Hoguvhi, 1915–16 (d)
Dr. Harry F. Noller, 1988–89 (a)
Dr. John H. Northrop, 1925–26, 1934–35 (d)
Dr. G.J.V. Nossal, 1967–68 (h)
Prof. Frederick G. Novy, 1934–35 (d)
Dr. Shosaku Numa, 1987–88 (a)
Dr. Paul Nurse, 1996–97
Dr. Ruth S. Nussenzwelg, 1982–83 (a)
Dr. Victor Nussenzweig, 1982–83 (a)
Dr. Christiane Nussleix-Volhard, 1990–91 (h)
Prof. George, H.F. Nuttall, 1912–13 (d)
Dr. Severo Ochoa, 1950–51 (a)
Dr. Lloyd J. Old, 1971–72, 1975–76 (h)
Dr. John Oliphant, 1943–44 (d)
Dr. Jean Oliver, 1944–45 (h)
Dr. Eric N. Olsen, 2002–03 (a)
Dr. Bert W. O'Malley, 1976–77 (h)
Dr. J.L. Oxcley, 1954–55 (h)
Dr. Eugene L. Opie, 1909–10, 1928–29, 1954–55 (d)
Dr. Moshe Oren, 2001–02
Dr. Stuart H. Orkin, 1987–88 (a)
Prof. Henry F. Osborn, 1911–12 (d)
Dr. Mary Jane Osborn, 1982–83 (h)
Dr. Thomas B. Osborne, 1910–11 (d)
Dr. Winthrop J.V. Osterhout, 1921–22, 1929–30 (d)
Dr. Norman R. Pace, 1995–96
Dr. George E. Palade, 1961–62 (a)
Dr. A.M. Pappenheimer, Jr., 1956–57, 1989–81 (a)
Dr. John R. Pappenheimer, 1965–66 (a)
Prof. Arthur B. Pardee, 1969–70 (h)
Dr. Edwards A. Park, 1938–39 (d)
Prof. W.H. Park, 1905–06 (d)
Prof. G.H. Parker, 1913–14 (d)
Dr. Stewart Paton, 1917–19 (d)
Dr. John R. Paul, 1942–43 (d)
Dr. L. Pauling, 1953–54 (h)
Dr. Nikola Pavletich, 2002–03 (a)
Dr. Francis W. Peabody, 1916–17 (d)

PROF. RICHARD M. PEARCE, 1909–10 (d)
DR. RAYMOND PEARL, 1921–22 (d)
DR. WILLIAM STANLEY PEART, 1977–78 (h)
DR. WILDER PENFIELD, 1936–37 (d)
DR. M.F. PERUTZ, 1967–68 (h)
DR. JOHN P. PETERS, 1937–38 (d)
DR. W.H. PETERSON, 1946–47 (d)
DR. DAVID C. PHILLIPS, 1970–71 (h)
DR. ERNST, P. PICK, 1929–30 (h)
DR. LUDWIG PICK, 1931–32 (d)
DR. GREGORY PINCUS, 1966–67 (d)
DR. CLEMENS PIRQUET, 1921–22 (d)
DR. COLIN PITENDRIGH, 1960–61 (h)
DR. ROBERT PITTS, 1952–53 (d)
DR. A. POLICARD, 1931–32 (h)
DR. THOMAS D. POLLARD, 2002–03 (a)
PROF. GEORGE J. POPJAK, 1969–70 (h)
DR. KETTH R. PORTER, 1955–56 (a)
PROF. RODNEY R. PORTER, 1969–70 (h)
DR. W.T. PORTER, 1906–07, 1917–19 (d)
DR. STANLEY B. PRUSINER, 1991–92 (h)
DR. PETER PRYCIAK, 1994–95
DR. MARK PTASHNE, 1973–74 (h)
DR. T.T. PUCK, 1958–59 (h)
DR. J.J. PUTNAM, 1911–12 (d)
DR. VINCENT R. RACANIELLO, 1991–92 (h)
DR. EFRAIM RACKER, 1955–56 (a)
DR. HERMANN RAHN, 1958–59 (h)
DR. CHARLES H. RAMMELKAMP, JR., 1955–56 (h)
DR. S. WALTER RANSON, 1937–37 (d)
DR. KENNETH B. RAPER, 1961–62 (h)
DR. TOM A. RAPOPORT, 2006–07 (a)
DR. ALEXANDER RICH, 1982–83 (a)
DR. ARNOLD R. RICH, 1946–57 (d)
PROF. ALFRED N. RICHARDS, 1920–21, 1934–35 (d)
DR. DICKINSON W. RICHARDS, 1943–44 (h)
PROF. THEODORE W. RICHARDS, 1911–12 (d)
DR. CURT P. RICHTER, 1942–43 (h)
DR. D. RITTENBERG, 1948–49 (d)

DR. THOMAS M. RIVERS, 1933–34 (d)
DR. WILLIAM ROBBINS, 1942–43(h)
DR. ELIZABETH J. ROBERTSON, 2005–06
DR. O.H. ROBERTSON, 1943–43 (d)
PROF. WILLIAM C. ROSE, 1934–35 (h)
PROF. ORA MENDELSOHN ROSEN, 1986–87 (a)
DR. M.J. ROSENAU, 1908–09 (d)
DR. RUSSELL ROSS, 1981–82 (a)
DR. JANET ROSSANT, 2001–02
DR. MICHAEL G. ROSSMANN, 1987–88 (a)
DR. JESSE ROTH, 1981–82 (a)
DR. JAMES E. ROTHMAN, 1990–91 (h)
DR. E.J.W. ROUGHTON, 1943–44 (h)
DR. PEYTON ROUS, 1935–36 (d)
DR. WALLACE P. ROWE, 1975–76 (h)
DR. GERALD M. RUBIN, 1987–88 (a)
DR. HARRY RUBIN, 1965–66 (h)
PROF. MAX RUBNER, 1912–13 (d)
DR. FRANK H. RUDDLE, 1973–74 (h)
DR. JOHN RUNNSTROM, 1950–51 (h)
DR. ERKKI RUOSLAHTI, 1988–89 (a)
MAJOR FREDERJCK F. RUSSELL, 1912–13 (d)
DR. GARY B. RUVKIN, 2003–04
DR. E.R. SABIN, 1915–16 (d)
DR. LEO SACHS, 1972–73 (h)
DR. RUTH SAGER, 1982–83 (h)
DR. BENGT SAMUELSSON, 1979–80 (h)
DR. WILBUR A. SAWYER, 1934–35 (d)
DR. HOWARD SCHACHMAN, 1972–73 (h)
PROF. E.A. SCHAFER, 1907–08 (d)
DR. GOTTFRIED SCHATZ, 1987–90 (h)
DR. ROBERT T. SCHIMKE, 1980–81 (h)
DR. MATTHEW D. SCHARFF, 1973–74 (a)
DR. HAROLD W. SCHERAGA, 1967–68 (h)
DR. BELA SCHICK, 1922–23 (h)
DR. ROBERT SCHLEIF, 1988–89 (a)
DR. JOSEPH SCHLESSINGER, 1993–94
DR. OSCAR SCHLOSS, 1924–25 (d)
DR. STUART F. SCHLOSSMAN, 1983–84 (h)
PROF. ADOLPH SCHMIDT, 1913–14 (d)
DR. CARL F. SCHMIDT, 1948–49 (h)

Dr. Knut Schmidt-Neilsen, 1962–63 (h)
Dr. Randy W. Schekman, 1994–95
Dr. Francis O. Schmitt, 1944–45 (h)
Dr. R. Schoeneheimer, 1936–37 (d)
Dr. P.E. Scholander, 1961–62 (h)
Dr. Stuart Lee Schreiber, 1955–96
Dr. Kathrin Schrick, 1994–95
Dr. Peter G. Schultz, 1999–00 (h)
Dr. Robert S. Schwartz, 1985–86 (h)
Dr. Joseph G. Sodroski, 2004–05
Dr. Nevin S. Srimshaw, 1962–63 (h)
Dr. William H. Sebrells, 1943–44 (h)
Prof. W.T. Sedgwick, 1911–12 (d)
Dr. Walter Seegers, 1951–52 (h)
Dr. J. Edwin Seegmiller, 1969–70 (h)
Dr. Michael Sela, 1971–72 (h)
Dr. Dennis J. Selkoe, 2003–04
Dr. Philip A. Shaffer, 1922–23 (d)
Dr. James A. Shannon, 1945–46 (h)
Dr. Lucy Shapiro, 1992–93
Dr. Phillip A. Sharp, 1985–86 (h)
Dr. George R. Stark, 1997–98
Dr. David Shemin, 1954–55 (h)
Dr. Henry C. Sherman, 1917–19 (d)
Dr. Charles Sherr, 2000–01
Dr. Richard Shope, 1935–36 (d)
Dr. Ephraim Shorr, 1954–55 (d)
Dr. Paul B. Sigler, 1999–00 (d)
Dr. Kai Simons, 1993–94
Dr. Robert L. Sinsheimer, 1968–69 (h)
Dr. E.C. Slater, 1970–71 (h)
Dr. G. Eliot Smith, 1930–31 (d)
Dr. Emil L. Smith, 1966–67 (h)
Dr. Homer W. Smith, 1939–40 (d)
Dr. Philip E. Smith, 1929–30 (d)
Prof. Theobald Smith, 1905–06 (d)
Dr. George D. Snell, 1978–79 (h)
Dr. Solomon H. Snyder, 1977–78 (h)
Dr. Louis Sokoloff, 1983–84 (h)
Dr. T.M. Sonneborn, 1948–49 (h)
Dr. S.P.L. Sorenson, 1924–25 (d)
Dr. Carl C. Speidel, 1940–41 (h)
Dr. Sol Spiegelman, 1968–69 (a)
Dr. Roger W. Sperry, 1966–67 (h)

Dr. Timothy A. Springer, 1993–94
Dr. William C. Stadie, 1941–42 (d)
Dr. Earl. R. Stadtman, 1969–70 (h)
Dr. Roger Stanier, 1959–60 (h)
Dr. Wendell Stanley, 1937–38 (d)
Dr. Earnest H. Starling, 1907–08 (d)
Dr. Isaac Starr, 1946–47 (h)
Dr. William H. Stein, 1956–57 (h)
Dr. Donald F. Steiner, 1982–83 (h)
Dr. Joan A. Steitz, 1984–85 (h)
Dr. Thomas E. Steitz, 1997–98
Dr. P. Stetson, 1927–28 (d)
Prof. George Stewart, 1912–13 (d)
Prof. Ch. Wardell Stiles, 1915–16 (d)
Dr. Bruce Stillman, 1992–93
Dr. C.R. Stockyard, 1921–22 (d)
Dr. Walter Straub, 1928–29 (h)
Dr. George L. Streeter, 1933–34 (h)
Dr. Jack L. Strominger, 1968–69 (h)
Dr. R.P. Strong, 1913–14 (d)
Dr. Lubert Stryer, 1991–92 (h)
Prof. Earl W. Sutherland, Jr., 1961–62 (d)
Prof. Homer F. Swift, 1919–20 (d)
Dr. W.W. Swingle, 1931–32 (d)
Dr. V.P. Sydenstricker, 1942–43 (h)
Dr. Albert Szent-Gyorgyi, 1938–39 (h)
Dr. Jack W. Szostak, 1997–98
Dr. Tadatsugu Taniguchi, 2005–06
Dr. W.H. Taliaferro, 1931–32 (d)
Prof. Alonzo E. Taylor, 1907–08 (d)
Dr. Howard M. Temin, 1973–74 (h)
Dr. Marc Tessier-Lavigne, 2002–03 (a)
Prof. W.S. Thayer, 1911–12 (d)
Dr. Hugo Theorell, 1965–66 (h)
Dr. Lewis Thomas, 1967–68 (a)
Dr. Shirley M. Tilghman, 1991–92 (h)
Dr. William S. Tillett, 1949–50 (h)
Dr. Arne Tiselius, 1939–40 (h)
Dr. Robert Tjian, 1994–95
Dr. A.R. Todd, 1951–52 (h)
Dr. Gordon M. Tomkins, 1972–73 (h)
Dr. Susumu Tonegawa, 1979–80 (h)

DR. ROGER Y. TSIEN, 2003–04
DR. SIDNEY UDENFRIEND, 1964–65 (h)
COLONEL F.P. UNDERHILL, 1917–19 (d)
DR. HANS USSING, 1963–64 (h)
DR. P. ROY VAGELOS, 1974–75 (h)
DR. DONALD D. VAN SLYKE, 1915–16 (d)
DR. HAROLD VARMUS, 1987–88 (a)
DR. ALEXANDER VARSHAVSKY, 2000–01
DR. MARTHA VAUGHAN, 1981–82 (h)
PROF. VICTOR C. VAUGHN, 1913–14 (d)
DR. GREGORY L. VERDINE, 2006–07 (a)
PROF. MAX VERWORN, 1911–12 (d)
PROF. CARL VOEGTLIN, 1919–20 (d)
DR. U.S. VON EULER, 1958–59 (h)
DR. ALEXANDER VON MURALT, 1947–48 (h)
PROF. CARL VON NOORDEN, 105–06 (d)
DR. SELMAN A. WAKSMAN, 1944–45 (d)
DR. GEORGE WALD, 1945–46 (h)
DR. JAN WALDENSTROM, 1960–61 (h)
PROF. THOMAS A. WALDMANN, 1986–87 (h)
DR. PETER WALKER, 1995–96
PROF. AUGUSTUS D. WALLER, 1913–14 (d)
DR. JAMES C. WANG, 1985–86 (h)
DR. JOSEF WARKANY, 1952–53 (h)
COLONEL STAFFORD L. WARREN, 1945–46 (h)
DR. ALFRED S. WARTHIN, 1917–19 (d)
DR. C.J. WATSON, 1948–49 (h)
DR. JOSEPH T. WEARN, 1939–40 (h)
DR. D.J. WEATHERALL, 1999–00 (a)
DR. H.H. WEBER, 1953–54 (d)
PROF. J. CLARENCE WEBSTER, 1905–06 (d)
DR. L.T. WEBSTER, 1931–32 (d)
DR. A. ASHLEY WEECH, 1938–39 (h)
DR. SILVIO WEIDMANN, 1965–66 (h)
DR. ROBERT A. WEINBERG, 1984–85 (h)
DR. HAROLD M. WEINTRAUB, 1983–84 (h)
DR. PAUL WEISS, 1958–59 (h)
IRVING L. WEISSMAN, 1989–90 (h)
DR. CHARLES WEISSMANN, 1981–82 (h)

DR. WILLIAM H. WELCH, 1915–16 (d)
DR. THOMAS H. WELLER, 1956–57 (h)
PROF. H. GIDEON WELLS, 1910–11 (d)
DR. K.F. WENCKEBACH, 1922–23 (d)
DR. GEORGE H. WHIPPLE, 1921–22 (d)
DR. ABRAHAM WHITE, 1947–48 (a)
DR. RAYMOND L. WHITE, 1984–85 (h)
DR. CARL J. WIGGERS, 1920–21, 1956–57 (d)
DR. V.B. WIGGLESWORTH, 1959–60 (h)
DR. DON C. WILEY, 1988–89 (a)
DR. CARROLL M. WILLIAMS, 1951–52 (h)
DR. LINSLEY R. WILLIAMS, 1917–19 (d)
DR. RICHARD WILLSTATTER, 1926–27 (d)
DR. EDMUND B. WILSON, 1906–07 (d)
DR. EDWIN B. WILSON, 1925–26 (d)
PROF. J. GORDON WILSON, 1917–19 (h)
DR. JEAN D. WILSON, 1983–84 (h)
DR. WILLIAM F. WINDLE, 1944–45 (h)
DR. F.R. WINTON, 1951–52 (h)
DR. MAXWELL M. WINTROBE, 1949–50 (d)
PROF. S.B. WOLFBACH, 1920–21 (d)
DR. HAROLD G. WOLFF, 1943–44 (d)
DR. HARLAND G. WOOD, 1949–50 (h)
DR. W. BARRY WOOD, JR., 1951–52 (d)
DR. WILLIAM B. WOOD, 1977–78 (h)
PROF. SIR MICHAEL F.A. WOODRUFF, 1970–71 (h)
DR. ROBERT B. WOODWARD, 1963–64 (h)
DR. R.T. WOODYATT, 1915–16 (d)
DR. D.W. WOOLLEY, 1945–46 (d)
SIR A LMROTH E. WRIGHT, 1906–07 (d)
DR. ROSALYN S. YALOW, 1966–67 (h)
DR. KEITH R. YAMAMOTO, 1995–96
PROF. ROBERT M. YERKES, 1917–19, 1935–36 (d)
DR. PAUL C. ZAMECNIK, 1959–60 (h)
DR. L. ZECHMEISTER, 1951–52 (h)
DR. NORTON D. ZINDER, 1966–67 (a)
DR. R.M. ZINKERNAGEL, 1993–94
PROF. HANS ZINSSER, 1914–15 (d)
DR. LEONARD I. ZON, 2006–07 (a)

ACTIVE MEMBERS

Dr. Geoffrey W. Abbott
Dr. John Abelson
Dr. Steven B. Abramson
Dr. George Acs
Dr. Janet Scott van Adelsberg
Dr. Alan Aderem
Dr. Aneel K. Aggarwal
Dr. Maria E. Aguero-Rosenfeld
Dr. Salah Al-Askari
Dr. Qais Al-Awqati
Dr. E. Jane Albert Hubbard
Dr. Michael H. Alderman
Dr. Emma G. Allen
Dr. David C. Allis
Dr. Frederick W. Alt
Dr. Blanche Pearl Alter
Dr. Burton M. Altura
Dr. Olaf Sparre Andersen
Dr. Giuseppe A. Andres
Dr. Ruth Angeletti
Dr. Reginald M. Archibald
Dr. Amir Askari
Dr. Manfred Auer
Dr. Arleen D. Auerbach
Dr. Peter A.M. Auld
Dr. Theodore W. Avruskin
Dr. Stephen M. Ayres
Dr. Efrain Charles Azmitia
Dr. Erika Bach
Dr. Richard A. Bader
Dr. Nick Baker
Dr. David S. Baldwin
Dr. Amiya Banerjee

Dr. Sachchidananda Banerjee
Dr. Arthur Bank
Dr. Norman Bank
Dr. Arlene Desiree Bardeguez
Dr. Margaret H. Baron
Dr. James J. Barondess
Dr. Jeremiah A. Barondess
Dr. Robert Bases
Dr. Claudio Basilico
Dr. Craig T. Basson
Dr. Olcay A. Batuman
Dr. Alexander G. Bearn
Dr. A. Robert Beck
Dr. Carl G. Becker
Dr. Bertrand M. Bell
Dr. Vivian Bellofatto
Dr. Baruj Benacerraf
Dr. Robert Benezra
Dr. Philip N. Benfey
Dr. Craig J. Benham
Dr. Bry Benjamin
Dr. Stephen J. Benkovic
Dr. Gordon D. Benson
Dr. Richard Beresford
Dr. Carole L. Berger
Dr. Lawrence Berger
Dr. Paul D. Berk
Dr. L.H. Bernstein
Dr. Peter Besmer
Dr. Rajesh M. Bhatnagar
Dr. Jahar Bhattacharya
Dr. Kailash C. Bhuyan
Dr. R.J. Bing

Dr. Barbara K. Birshtein
Dr. David Bishop
Dr. Elizabeth H. Blackburn
Dr. John Anthony Blaho
Dr. William S. Blaner
Dr. Carl P. Blobel
Dr. Gunter K.J. Blobel
Dr. Bernard H. Boal
Dr. Richard Steven Bockman
Dr. Bruce I. Bogart
Dr. Alfred J. Bollet
Dr. Robert M. Bookchin
Dr. William Borkowsky
Dr. Adele L. Boskey
Dr. Robert J. Boylan
Dr. Leon Bradlow
Dr. Jo Anne Brasel
Dr. Goodwin M. Breinin
Dr. Esther Breslow
Dr. Jan L. Breslow
Dr. Robin W. Briehl
Dr. Jonathan S. Bromberg
Dr. Felix Bronner
Dr. Mark S. Brower
Dr. Steven T. Brower
Dr. Clinton D. Brown
Dr. William Ted Brown
Dr. Doris J. Bucher
Dr. Jochen Buck
Dr. Joseph A. Buda
Dr. Hannes E. Buelow
Dr. Steven Burden
Dr. Robert D. Burk
Dr. Joseph Buxbaum
Dr. Robert Cancro
Dr. Charles R. Cantor
Dr. Joseph M. Capasso
Dr. Marian Carlson
Dr. Peter W. Carmel
Dr. Marc G. Caron
Dr. Hugh J. Carroll
Dr. Patrizia Casaccia-Bonnefil
Dr. Arturo Casadevall
Dr. Joan I. Casey
Dr. Christine Karen Cassel

Dr. Veronica M. Catanese
Dr. Daniel F. Catanzaro
Dr. Peter Cervoni
Dr. Raju S.K. Chaganti
Dr. Brian Trevor Chait
Dr. Michael J. Chamberlin
Dr. Robert Warner Chambers
Dr. W.Y. Chan
Dr. Fred Chang
Dr. Moses Victor Chao
Dr. Mitchell Charap
Dr. Norman E. Chase
Dr. Herbert S. Chase, Jr.
Dr. Mark R. Chassin
Dr. Tie Chen
Dr. Selina Chen Kiang
Dr. Leonard Chess
Dr. David Chi
Dr. Geoffrey J. Childs
Dr. Francis P. Chinard
Dr. Yong Sung Choi
Dr. Purnell W. Choppin
Dr. Ting-Chao Chou
Dr. Dennis J. Cleri
Dr. William L. Cleveland
Dr. David E. Cobrinik
Dr. Gerald Cohen
Dr. Ira S. Cohen
Dr. Paula E. Cohen
Dr. Randolph P. Cole
Dr. Morton Coleman
Dr. Barry S. Coller
Dr. John Condeelis
Dr. Lawrence A. Cone
Dr. Stuart D. Cook
Dr. Norman S. Cooper
Dr. Jack M. Cooperman
Dr. Lucien J. Cote
Dr. David Cowburn
Dr. David Cowen
Dr. Eva Brown Cramer
Dr. Bruce N. Cronstein
Dr. Frederick R. Cross
Dr. George A.M. Cross
Dr. Kathryn L. Crossin

Dr. Bruce Cunningham
Dr. Charlotte Cunningham-Rundles
Dr. Susanna Cunningham-Rundles
Dr. Samuel J. Danishefsky
Dr. James E. Darnell
Dr. Robert B. Darnell
Dr. Seth Andrew Darst
Dr. John R. David
Dr. Terry Francis Davies
Dr. Jean Davignon
Dr. Robert P. Davis
Dr. Paul F. De Gara
Dr. Robert H. DeBellis
Dr. Vittorio Defendi
Dr. Thomas J. Degnan
Dr. Ralph B. Dell
Dr. John R. Denton
Dr. Robert J. Desnick
Dr. Claude Desplan
Dr. Dickson D. Despommier
Dr. Patricia A. Detmers
Dr. Thomas Detwiler
Dr. Bhanumas Dharmgrongartama
Dr. Madhav V. Dhodapkar
Dr. Salvatore Di Mauro
Dr. Joseph R. Di Palma
Dr. Betty Diamond
Dr. Adam P. Dicker
Dr. Alberto DiDonato
Dr. Alexandra B. Dimich
Dr. Ann M. Dnistrian
Dr. Jane Dodd
Dr. Alvin M. Donnenfeld
Dr. Burton S. Dornfest
Dr. Dale Dorsett
Dr. Steven D. Douglas
Dr. Peter Clowes Dowling
Dr. Burton Drayer
Dr. Paul Dreizen
Dr. Michael J. Droller
Dr. Lewis M. Drusin
Dr. Ronald E. Drusin
Dr. Bo Dupont
Dr. Michael L. Dustin
Dr. Jonathan Dworkin

Dr. Brian D. Dynlacht
Dr. Laurel A. Eckhardt
Dr. Gerald M. Edelman
Dr. Isidore S. Edelman
Dr. Winfried Edelmann
Dr. Richard L. Edelson
Dr. Mark L. Edwards
Dr. Hans J. Eggers
Dr. Herman N. Eisen
Dr. Robert P. Eisinger
Dr. David Eliezer
Dr. Nathan Ames Ellis
Dr. Peter Elsbach
Dr. Samuel Elster
Dr. Scott W. Emmons
Dr. Doruk Erkan
Dr. Bernard F. Erlanger
Dr. Alice L. Erwin
Dr. Diane Esposito
Dr. Thomas J. Fahey
Dr. Marianne C. Fahs
Dr. David H. Farb
Dr. Mehdi Farhangi
Dr. Nunzia S. Fatica
Dr. Daniel S. Feldman
Dr. Marie Theresa Filbin
Dr. Arthur D. Finck
Dr. Stuart Firestein
Dr. Vincent Fischetti
Dr. Donald Fischman
Dr. Paul B. Fisher
Dr. Glenn I. Fishman
Dr. David Fitch
Dr. Kathleen M. Foley
Dr. David A. Foster
Dr. Arthur C. Fox
Dr. Steven Fox
Dr. Carl Frasch
Dr. Manfred Frasch
Dr. Blair A. Fraser
Dr. Leonard Paul Freedman
Dr. Michael L. Freedman
Dr. Alan H. Friedman
Dr. Eli A. Friedman
Dr. Jeffrey M. Friedman

Dr. Stanley Friedman
Dr. Steven Fruchtman
Dr. Elaine Fuchs
Dr. Hiroyoshi Fujita
Dr. Hironori Funabiki
Dr. Robert F. Furchgott
Dr. Henry Furneaux
Dr. Mark E. Furth
Dr. Yasuhiro Furuichi
Dr. Terry Gaasterland
Dr. Jacques L. Gabrilove
Dr. Janice L. Gabrilove
Dr. David C. Gadsby
Dr. W. Einar Gall
Dr. Bruce Ganem
Dr. Martin Gardy
Dr. Frederick Gates, III
Dr. Mario Gaudino
Dr. Ulrike Gaul
Dr. Lester M. Geller
Dr. Jeremiah M. Gelles
Dr. Donald A. Gerber
Dr. James German, III
Dr. Michael D. Gershon
Dr. Elizabeth C. Gerst
Dr. Menard M. Gertler
Dr. Ranajeet Ghose
Dr. Ranjeet Ghose
Dr. Allan Gibofsky
Dr. Charles D. Gilbert
Dr. Alfred G. Gilman
Dr. Charles Gilvarg
Dr. David Y. Gin
Dr. Henry Ginsberg
Dr. Alan Gintzler
Dr. Mark E. Girvin
Dr. Warren Glaser
Dr. Ephraim Glassman
Dr. David L. Globus
Dr. Henry P. Godfrey
Dr. G. Nigel Godson
Dr. Leslie I. Gold
Dr. Ira J. Goldberg
Dr. Marcia B. Goldberg
Dr. Lewis Goldfrank

Dr. Laura T. Goldsmith
Dr. Gideon Goldstein
Dr. Harris Goldstein
Dr. Marvin H. Goldstein
Dr. Charles Gonzalez
Dr. Emil C. Gotschlich
Dr. R.F. Grady
Dr. Angela Granelli-Piperno
Dr. Michael R. Green
Dr. Olga Greengard
Dr. R.A. Gregory
Dr. Roger L. Greif
Dr. Ira Greifer
Dr. Anthony J. Grieco
Dr. Randall B. Griepp
Dr. Joel Grinker
Dr. David Grob
Dr. Steven S. Gross
Dr. Lionel Grossbard
Dr. Vladimir Grubor
Dr. Melvin M. Grumbach
Dr. Joseph J. Guarneri
Dr. Lorraine J. Gudas
Dr. Giancarlo Guideri
Dr. Guido Guidotti
Dr. Jose Guillem
Dr. Subhash Gulati
Dr. Susan Hadley
Dr. Jack W.C. Hagstrom
Dr. Kathleen A. Haines
Dr. David P. Hajjar
Dr. Katherine A. Hajjar
Dr. Mimi Halpern
Dr. James B. Hamilton
Dr. Leonard D. Hamilton
Dr. Scott M. Hammer
Dr. Ulrich Hammerling
Dr. Meirong Hao
Dr. Michael B. Harris
Dr. Franz-Ulrich Hartl
Dr. George A. Hashim
Dr. Sami Hashim
Dr. Victor Hatcher
Dr. Mary Elizabeth Hatten
Dr. A. Daniel Hauser

Dr. Richard Hawkins
Dr. R. Scott Hawley
Dr. Arthur Pearman Hays
Dr. Richard M. Hays
Dr. John Healey
Dr. Michael Heidelberger
Dr. Nathaniel Heintz
Dr. Samuel Hellman
Dr. Ali Hemmati-Brivanlou
Dr. John J. Hemperly
Dr. Walter L. Henley
Dr. Michael V. Herman
Dr. Margaret W. Hilgartner
Dr. Jules Hirsch
Dr. Kurt Hirschhorn
Dr. Rochelle Hirschhorn
Dr. David I. Hirsh
Dr. David D. Ho
Dr. John L. Ho
Dr. Oliver Hobert
Dr. Paul Hochstein
Dr. Raymond F. Holden, Jr.
Dr. Robert S. Holzman
Dr. Curt M. Horvath
Dr. Susan Horwitz
Dr. Alan N. Houghton
Dr. Ling-Ling Hsieh
Dr. Howard H.T. Hsu
Dr. Jerard Hurwitz
Dr. Mahmood Hussain
Dr. Antonio Iavarone
Dr. Laura Inselman
Dr. Harry L. Ioachim
Dr. Luis M. Isola
Dr. Fuyuki Iwasa
Dr. Richard W. Jackson
Dr. William R. Jacobs, Jr.
Dr. Sheldon Jacobson
Dr. Rudolf Jaenisch
Dr. Ernst R. Jaffe
Dr. Herbert Jaffe
Dr. Gary D. James
Dr. Charles I. Jarowski
Dr. Maria Jasin
Dr. Jamshid Javid

Dr. Daniel C. Javitt
Dr. Norman B. Javitt
Dr. S. Michal Jazwinski
Dr. Rolf Jessberger
Dr. Thomas Jessell
Dr. Sulin Jiang
Dr. Edward M. Johnson
Dr. Saran Jonas
Dr. Richard Jove
Dr. Alexandra L. Joyner
Dr. David B. Kaback
Dr. Ronald Kaback
Dr. Lawrence J. Kagan
Dr. Melvin Kahn
Dr. Thomas Kahn
Dr. Alfred J. Kaltman
Dr. Mikio Kamiyama
Dr. Sandra Kammerman
Dr. Eric R. Kandel
Dr. Yoshinobu Kanno
Dr. Gilla Kaplan
Dr. Tarun M. Kapoor
Dr. Attallah Kappas
Dr. Arthur Karanas
Dr. Maria Karayiorgou
Dr. Arthur Karlin
Dr. Stuart S. Kassan
Dr. Michael Katz
Dr. Herbert J. Kayden
Dr. Gordon I. Kaye
Dr. Scott Neal Keeney
Dr. Hanna Kelker
Dr. Gordon M. Keller
Dr. Patrick Kelly
Dr. Thomas J. Kelly
Dr. Anna Marie Kenney
Dr. Michael Christopher Keogh
Dr. Leo Kesner
Dr. Richard H. Kessin
Dr. Yutaka Kikkawa
Dr. Yoon Berm Kim
Dr. Tom Kindt
Dr. Donald West King
Dr. David W. Kinne
Dr. Jan Kitajewski

Dr. David G. Klapper
Dr. Jack Klatell
Dr. Janis V. Klavins
Dr. David Kleinberg
Dr. Jerome L. Knittle
Dr. Andrew Koff
Dr. Kiyomi Koizumi
Dr. Richard Neal Kolesnick
Dr. Edwin H. Kolodny
Dr. Masakazu Konishi
Dr. Arthur Kornberg
Dr. Ione A. Kourides
Dr. Thomas R. Kozel
Dr. Rosemary Kraemer
Dr. Alfred N. Krauss
Dr. Robert Krauss
Dr. Mary Jeanne Kreek
Dr. Gert Kreibich
Dr. Susan E. Krown
Dr. James G. Krueger
Dr. Terry Ann Krulwich
Dr. Edward J. Kuchinskas
Dr. Friedrich Kueppers
Dr. Ashok B. Kulkarni
Dr. Ashok Kumar
Dr. Henn Kutt
Dr. Sau-Ping Kwan
Dr. Michael P. La Quaglia
Dr. Elizabeth H. Lacy
Dr. Robert G. Lahita
Dr. Emmanuel M. Landau
Dr. Philip John Landrigan
Dr. Frank R. Landsberger
Dr. Daniel Lane
Dr. Titia de Lange
Dr. Philip Lanzkowsky
Dr. John H. Laragh
Dr. Etienne Y. Lasfargues
Dr. S.E. Lasker
Dr. John Lattimer
Dr. Paul B. Lazarow
Dr. Robert A. Lazzarini
Dr. Mark Gabriel Lebwohl
Dr. Mathew Lee
Dr. Stanley L. Lee

Dr. Sylvia Lee-Huang
Dr. Albert M. Lefkovits
Dr. Robert J. Lefkowitz
Dr. Thomas J.A. Lehman
Dr. Ruth Lehmann
Dr. David Lehr
Dr. Rudolph L. Leibel
Dr. Stanislas Leibler
Dr. Edgar Leifer
Dr. Leslie A. Leinwand
Dr. Neal S. LeLeiko
Dr. Philip L. Leopold
Dr. Gerson T. Lesser
Dr. Stanley M. Levenson
Dr. Roberto Levi
Dr. Lonny Ray Levin
Dr. Robert A. Levine
Dr. Harvey M. Levy
Dr. Gloria C. Li
Dr. Jonathan D. Licht
Dr. Charles S. Lieber
Dr. Seymour Lieberman
Dr. Christopher D. Lima
Dr. Michael R. Linden
Dr. Susan L. Lindquist
Dr. Thierry Jean Lints
Dr. Hsiou-Chi Liou
Dr. George Lipkin
Dr. Martin Lipkin
Dr. Dan R. Littman
Dr. Arthur H. Livermore
Dr. Arthur H. Livermore
Dr. Philip Livingston
Dr. Rodolfo Llinas
Dr. John N. Loeb
Dr. John D. Loike
Dr. Irving M. London
Dr. Jerome Lowenstein
Dr. Bingwei Lu
Dr. A. Leonard Luhby
Dr. Daria Anne Luisi
Dr. Daniel S. Lukas
Dr. Roderick MacKinnon
Dr. Marcelo O. Magnasco
Dr. Richard J. Mahler

Dr. Umadas Maitra
Dr. Craig C. Malbon
Dr. Sridhar Mani
Dr. Richard S. Mann
Dr. James M. Manning
Dr. Aaron J. Marcus
Dr. David L. Marcus
Dr. Philip I. Marcus
Dr. Robert F. Margolskee
Dr. Catherine T. Marino
Dr. Paul A. Marks
Dr. Philippa Marrack
Dr. Lorraine Marsh
Dr. Bento Mascarenhas
Dr. Joan Massague
Dr. Leonard M. Mattes
Dr. Robert Matz
Dr. Frederick R. Maxfield
Dr. Lester T. May
Dr. Klaus Mayer
Dr. Lloyd F. Mayer
Dr. Vincent J. McAuliffe
Dr. Kenneth S. McCarty
Dr. Maclyn McCarty
Dr. Patricia McCormack
Dr. Julianne Imperato McGinley
Dr. Paul R. McHugh
Dr. Robert McVie
Dr. John G. Mears
Dr. U. Thomas Meier
Dr. Edward Meilman
Dr. John Mekalanos
Dr. Myron R. Melamed
Dr. Robert B. Mellins
Dr. Anant K. Menon
Dr. Jacques P. Merab
Dr. Maureen Mesis-Sanz
Dr. Tamar Michaeli
Dr. Joseph Michl
Dr. Donna Mildvan
Dr. Frederick Miller
Dr. Myron Miller
Dr. Donald R. Mills
Dr. C. Richard Minick
Dr. Timothy Mitchison

Dr. Marek Mlodzik
Dr. Peter Model
Dr. Carlos A. Molina
Dr. Peter Mombaerts
Dr. Carl Monder
Dr. Malcolm A.S. Moore
Dr. Brian L.G. Morgan
Dr. Gilda Morillocucci
Dr. Thomas Morris
Dr. John Morrison
Dr. John Henry Morrison
Dr. Anne Moscona
Dr. J. Anthony Movshon
Dr. Thomas W. Muir
Dr. William A. Muller
Dr. Christian Munz
Dr. Ben A. Murray
Dr. Ralph L. Nachman
Dr. Ronald L. Nagel
Dr. Harris M. Nagler
Dr. Tatsuji Namba
Dr. Kim Nasmyth
Dr. Stanley G. Nathenson
Dr. Clayton L. Natta
Dr. Brian A. Naughton
Dr. Lewis S. Nelson
Dr. Maria New
Dr. Carol Shaw Newlon
Dr. Hanh T. Nguyen
Dr. Warren W. Nichols
Dr. Julian Niemetz
Dr. Lee Ann Niswander
Dr. Robert Nolan
Dr. Allen J. Norin
Dr. Alison Jane North
Dr. Richard Novick
Dr. Phyliss M. Novikoff
Dr. Neil J. Nusbaum
Dr. Michel Nussenzweig
Dr. Ruth Nussenzweig
Dr. Victor Nussenzweig
Dr. Christiane Nusslein-Volhard
Dr. Timothy P. O'Connor
Dr. Manuel Ochoa
Dr. Severo Ochoa

Dr. Michael O'Donnell
Dr. Herbert F. Oettgen
Dr. Michiko Okamoto
Dr. Charles Warren Olanow
Dr. Eric N. Olson
Dr. Richard O'Reilly
Dr. Irwin Oreskes
Dr. Ernest V. Orsi
Dr. Louis Ortega
Dr. Roman Osman
Dr. Zoltan Ovary
Dr. Norbert I.A. Overweg
Dr. John Owen
Dr. Harvey L. Ozer
Dr. Paolo A. Paciucci
Dr. Stephen A. Paget
Dr. Elisabeth Paietta
Dr. Peter Palese
Dr. Arthur G. Palmer
Dr. Pier Paolo Pandolfi
Dr. William Pao
Dr. Nina Papavasiliou
Dr. George D. Pappas
Dr. Chae Gyu Park
Dr. Thomas S. Parker
Dr. Fiorenzo Paronetto
Dr. Pedro Pasik
Dr. Mark W. Pasmantier
Dr. Gavril W. Pasternak
Dr. Dinshaw J. Patel
Dr. Christian C. Patrick
Dr. Nikola Pavletich
Dr. Nikola P. Pavletich
Dr. William Stanley Peart
Dr. Ellinor B. Peerschke
Dr. Audrey S. Penn
Dr. Robert G. Pergolizzi
Dr. Benvenuto G. Pernis
Dr. Demetrius Pertsemlidis
Dr. Richard G. Pestell
Dr. Sidney Pestka
Dr. Donald W. Pfaff
Dr. Lawrence M. Pfeffer
Dr. Karl H. Pfenninger
Dr. Mark R. Philips

Dr. Robert A. Phillips
Dr. Julia M. Phillips-Quagliata
Dr. Michael Pillinger
Dr. Matthew Pincus
Dr. Liise-anne Pirofski
Dr. Xavier Pi-Sunyer
Dr. Steven Maurice Podos
Dr. Beatriz G.T. Pogo
Dr. Jeffrey W. Pollard
Dr. Thomas Pollard
Dr. Roberta Pollock
Dr. Bruce Polsky
Dr. Melissa Pope
Dr. Jerome G. Porush
Dr. Jerome B. Posner
Dr. Vinayaka R. Prasad
Dr. Gregory Prelich
Dr. Stanley B. Prusiner
Dr. Michael B. Prystowsky
Dr. Ellen Pure
Dr. Dominick P. Purpura
Dr. Charles C. Query
Dr. Vincent R. Racaniello
Dr. Kanti R. Rai
Dr. Francesco Ramirez
Dr. Jacob Rand
Dr. Gwendalyn J. Randolph
Dr. Helen M. Ranney
Dr. Aaron Rausen
Dr. Jeffrey V. Ravetch
Dr. Lawrence W. Raymond
Dr. Colvin M. Redman
Dr. George E. Reed
Dr. George N. Reeke, Jr.
Dr. Gabrielle H. Reem
Dr. Joan Reibman
Dr. Rosemary Reinke
Dr. Marilyn D. Resh
Dr. Charles M. Rice
Dr. Ronald F. Rieder
Dr. Richard A. Rifkind
Dr. Lynn S. Ripley
Dr. Hugh D. Robertson
Dr. Alan G. Robinson
Dr. Enrique Rodriguez-Boulan

Dr. Robert G. Roeder
Dr. Charles E. Rogler
Dr. William N. Rom
Dr. David Ron
Dr. Herbert G. Rose
Dr. John F. Rosen
Dr. Neal Rosen
Dr. Barry Rosenstein
Dr. William Rosner
Dr. David B. Roth
Dr. Jesse Roth
Dr. Sheldon P. Rothenberg
Dr. James E. Rothman
Dr. Paul Rothman
Dr. Rodney Rothstein
Dr. Michael P. Rout
Dr. Lewis P. Rowland
Dr. Paul Royce
Dr. S. Jaime Rozovski
Dr. Robert J. Ruben
Dr. B.A. Rubin
Dr. Jaime S. Rubin
Dr. Arye Rubinstein
Dr. Marjorie Russel
Dr. Hyung Don Ryoo
Dr. David Sabatini
Dr. David B. Sachar
Dr. Thomas P. Sakmar
Dr. Gerald Salen
Dr. Letty G.M. Salentijn
Dr. Andrej Sali
Dr. Irving Salit
Dr. Jane E. Salmon
Dr. Milton R.J. Salton
Dr. Herbert H. Samuels
Dr. Chris Sander
Dr. Regina M. Santella
Dr. Nanette F. Santoro
Dr. Shigeru Sassa
Dr. Birgit Satir
Dr. Peter Satir
Dr. Bernard V. Sauter
Dr. David G. Savage
Dr. Brij B. Saxena
Dr. Matthew D. Scharff

Dr. Gottfried Schatz
Dr. David A. Scheinberg
Dr. Barbara M. Scher
Dr. Jose V. Scher
Dr. William Scher
Dr. Lawrence Scherr
Dr. Alexander Schier
Dr. Peter B. Schiff
Dr. Gerald Schiffman
Dr. Joseph Schlessinger
Dr. Vern L. Schramm
Dr. Nicole Schreiber-Agus
Dr. Ulrich K. Schubart
Dr. Edward H. Schuchman
Dr. Steward Schuman
Dr. Rise Schwab
Dr. Ernest Schwartz
Dr. Irving L. Schwartz
Dr. James H. Schwartz
Dr. Steven Schwartz
Dr. John J. Sciarra
Dr. Leonard J. Sciorra
Dr. Sheldon J. Segal
Dr. Pravinkumar B. Sehgal
Dr. Mindell Seidlin
Dr. Irving Seidman
Dr. Samuel Seifter
Dr. Michael Sela
Dr. Stephen J. Seligman
Dr. Licia Selleri
Dr. Vijayasaradhi Setaluri
Dr. Bridget Shafit-Zagardo
Dr. David Shafritz
Dr. Shai Shaham
Dr. Herman S. Shapiro
Dr. Lawrence Shapiro
Dr. Lucille Shapiro
Dr. Warren B. Shapiro
Dr. Michael P. Sheetz
Dr. Michael B. Sheffery
Dr. Michael Shelanski
Dr. Dennis Shields
Dr. Moshe Shike
Dr. Abraham Shulman
Dr. M.A.Q. Siddiqui

Dr. Kohji Uchizono
Dr. Hiroshi Ueno
Dr. Jonathan W. Uhr
Dr. John E. Ultmann
Dr. Jay C. Unkeless
Dr. Mark L. Urken
Dr. Fred T. Valentine
Dr. Mario Vassalle
Dr. Mary A. Vogler
Dr. Leslie B. Vosshall
Dr. Bernard W. Wagner
Dr. John A. Wagner
Dr. Lila A. Wallis
Dr. Da-neng Wang
Dr. Jen C. Wang
Dr. Yu-Hwa Eugenia Wang
Dr. Peter E. Warburton
Dr. Peter Ward
Dr. Jonathan R. Warner
Dr. Paul M. Wassarman
Dr. Sylvia Wassertheil-Smoller
Dr. Samuel Waxman
Dr. Annemarie Weber
Dr. Thomas Weber
Dr. Richard Weil, III
Dr. Judah Weinberger
Dr. Harel Weinstein
Dr. Gerson Weiss
Dr. Irving L. Weissman
Dr. Gerald Weissmann
Dr. Babette B. Weksler
Dr. Marc E. Weksler

Dr. Daniel Wellner
Dr. Abraham G. White
Dr. Robert Wieder
Dr. Martin Wiedmann
Dr. Peter H. Wiernik
Dr. Torsten N. Wiesel
Dr. John E. Wilson
Dr. Thomas Wisniewski
Dr. Murray Wittner
Dr. David Wolf
Dr. Savio L.C. Woo
Dr. Hao Wu
Dr. George Yancopoulos
Dr. Seiichi Yasumura
Dr. Yusuf Yazici
Dr. Ghassan Yehia
Dr. Deborah Yelon
Dr. Peter I. Yi
Dr. Michael W. Young
Dr. Stuart H. Young
Dr. Michael S. Yuan
Dr. Rafael Yuste
Dr. John Zabriskie
Dr. Ralph Zalusky
Dr. Ming-Ming Zhou
Dr. Liang Zhu
Dr. Karen S. Zier
Dr. Norton D. Zinder
Dr. R.M. Zinkernagel
Dr. Geoffrey Louis Zubay
Dr. Dorothea Zucker-Franklin
Dr. Arturo Zychlinsky